Computational Physics and Cellular Automata

Workshop on
Computational Physics And Cellular Automata

Ouro Preto, Brazil   August 8-11, 1989

Editors:
A. Pires, D.P. Landau & H. Herrmann
Universidade Federal de Minas Gerais
Brazil

World Scientific
Singapore • New Jersey • London • Hong Kong

Published by

World Scientific Publishing Co Pte Ltd
5 Toh Tuck Link, Singapore 596224
USA office: 27 Warren Street, Suite 401-402, Hackensack, NJ 07601
UK office: 57 Shelton Street, Covent Garden, London WC2H 9HE

British Library Cataloguing-in-Publication Data
A catalogue record for this book is available from the British Library.

COMPUTATIONAL PHYSICS AND CELLULAR AUTOMATA

ISBN-13 978-981-02-0074-9
ISBN-10 981-02-0074-9

PREFACE

The present book is a collection of the lectures and posters presented at the Workshop on Computational Physics and Cellular Automata held at Ouro Preto, Brazil, August 8 to 11, 1989, where experts and young researchers came together to discuss computational physics.

The material presented in this book will be of interest to graduate students and researchers engaged in studies of application of computation to physics.

The Conference facilities were provided by the Escola de Minas da Universidade Federal de Ouro Preto. Financial support for the Workshop was received from the Conselho Nacional de Desenvolvimento Científico e Tecnológico (CNPq), Financiadora de Estudos e Projetos (FINEP) and IBM Brazil. We gratefully acknowledge the financial support of these Institutions.

We would also like to express our thanks to Prof. Cristovam Paes de Oliveira, rector of the Universidade Federal de Ouro Preto and to Dr. J. A. Plascak, Dr. J. K. L. da Silva, Dr. M. E. Gouvea, Dr. S. L. Talim, Mr. C. E. H. Sá Motta, and all the staff members of the Escola de Minas and of the Physics Department (Universidade Federal de Minas Gerais) for their help in organizing the workshop.

Finally we wish to thank all the participants for their contribution to the success of the workshop.

Prof. Antonio Pires
Physics Department
Universidade Federal de Minas Gerais

CONTENTS

Computational Physics and Cellular Automata

SPIN DYNAMICS SIMULATIONS OF EXCITATIONS IN SPIN CHAINS

D. P. Landau, R. W. Gerling[1], and M. S. S. Challa[2]
Center for Simulational Physics
The University of Georgia
Athens, GA 30602 U.S.A.

1. ABSTRACT

Spin dynamics methods are used to study the time dependent behavior of classical XY model spin chains in both zero and non-zero fields. Linear (spin waves) excitations are observed along with non-linear (multi-spin waves and solitons) modes.

2. INTRODUCTION

Magnetism in one dimension has long been a topic of great interest[1]. Physical systems of very weakly coupled magnetic chains can be readily fabricated and have been shown to have pseudo-one-dimensional behavior over wide ranges of temperature. Low temperature excitations are expected to be simple (linear) spin waves which can be easily measured by neutron scattering. Of great recent interest are one-dimensional magnets in a symmetry breaking field which are also expected to show non-linear excitations-solitons[2].

Central peak behavior in the inelastic neutron scattering results for $CsNiF_3$ in a magnetic field[3-4] first suggested that solitons were present in this system. A theoretical examination of a simple classical easy plane ferromagnet was made by mapping it, in the low temperature limit, onto the sine-Gordon model[5-7] whose dynamical prop-

[1]Permanent address: Institut für Theoretische Physik
Universität Erlangen-Nürnberg
8520 Erlangen, W. Germany
[2]Present Address: Physics Department
Virginia Commonwealth University
Richmond, VA 23284 U.S.A.

erties are well known. The comparison between theory and experiment was ambiguous because of the differences in the physical and theoretical Hamiltonians and because of the uncertainty in the range of validity of the theory. Solitons in antiferromagnetic chains, e.g. TMMC, have also been studied experimentally[8] and theoretically[9–10] but these systems are also only partially understood. We therefore chose to study the simple, classical XY-chain in order to assess the applicability of the theory.

The Hamiltonian for this model is

$$\mathcal{H} = -J\sum_i (S_i^x S_{i+1}^x + S_i^y S_{i+1}^y) - h\sum_i S_i^x \tag{1}$$

where $\vec{S}_i$ is a three component vector of unit length. The exchange constant J may either be positive (ferromagnetic) or negative (antiferromagnetic) and h is a uniform symmetry breaking field. The types of solitons which may be present are shown in Fig. 1. Previous Monte Carlo simulations have demonstrated the existence of 2π-solitons in the ferromagnetic case and type A and type B "π-like" solitons in the antiferromagnetic case. We have previously reported various aspects of this study elsewhere,[11–15] and a more extensive background can be found in these papers.

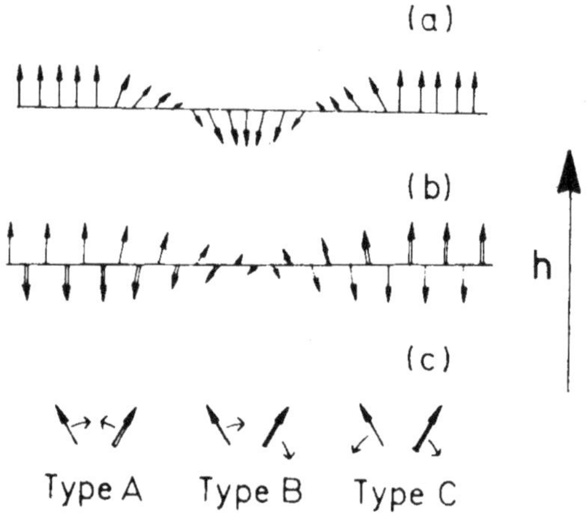

Fig. 1. Solitons in the XY-chain: (a) 2π-soliton in the ferromagnet. The spins rotate in the XY-plane through angle 2π; (b) π-soliton in the anisotropic antiferromagnet. The spins on each sublattice rotate in the XY-plane through angle π; (c) "π-like" solitons in the isotropic antiferromagnet. The spins are in a spin-flop state and each type of soliton is produced by allowing the sublattices to rotate in the XY-plane in the directions shown until they are interchanged.

3. METHOD

Equilibrium spin configurations for chains of length N=20,000 were generated using an importance sampling Monte Carlo method and used as starting configurations for the spin dynamics calculation. The equation of motion for each spin is

$$\dot{\vec{S}}_i = \vec{S}_i \times [J(S^x_{i+1} + S^x_{i-1})\vec{e_x} + J(S^y_{i+1} + S^y_{i-1})\vec{e_y} + h\vec{e_x}] \tag{2}$$

where $\vec{e_x}$ and $\vec{e_y}$ are unit vectors. These coupled, non-linear equations of motion were integrated on a Cyber 205 using a vectorized, 4th order predictor-corrector method with time integration interval $\Delta = 0.01J^{-1}$ out to a maximum time of $t_{max} = 100J^{-1}$. From each set of time dependencies we extracted space-displaced, time-displaced correlation functions:

$$< S^\alpha_i(0)S^\alpha_{i+r}(t) >= \frac{1}{N}\sum_{i=1}^{N} S^\alpha_i(0)S^\alpha_{i+r}(t) - < S^\alpha_i(0) >^2 \tag{3}$$

where r is the distance between sites (in units of the lattice constant) and $\alpha = x, y, z$ gives the polarization. The neutron scattering cross-section $S_\alpha(q,\omega)$ was then determined by the double Fourier transform

$$S_\alpha(q,\omega) = \sqrt{\frac{2}{\pi}} \int_0^{t_{max}} \sum_{r=1}^{r_{max}} \cos(qr)\cos(\omega t) < S^\alpha_i(0)S^\alpha_{i+r}(t) > e^{-\frac{1}{2}(r\cdot\delta r)^2} e^{-\frac{1}{2}(t\cdot\delta t)^2} dt \tag{4}$$

where the two Gaussians are included to reduce the effects of cutoffs on the transforms thus producing equivalent resolution limitations to those found in real experiments. All data presented here represent averages over 20 different initial configurations.

4. RESULTS

4.1 Ferromagnet ($J > 0$)

Extensive data now exist for the case of ferromagnetic exchange ($J > 0$). At low temperatures $S_z(q,\omega)$ shows single spin wave peaks, see Fig. 2, which can be easily fitted by resolution broadened Lorentzian lineshapes. $S_y(q,\omega)$ also shows pronounced spin-wave peaks; in Fig. 3 we show how these broaden and shift to lower frequency as the temperature increases. With increasing temperature a clear, Gaussian central peak develops indicating the presence of solitons.

The q-dependence of $S_y(q,\omega)$ is shown in Fig. 4. The single spin-wave peaks move up in frequency with larger q, and the central peak decreases in intensity and

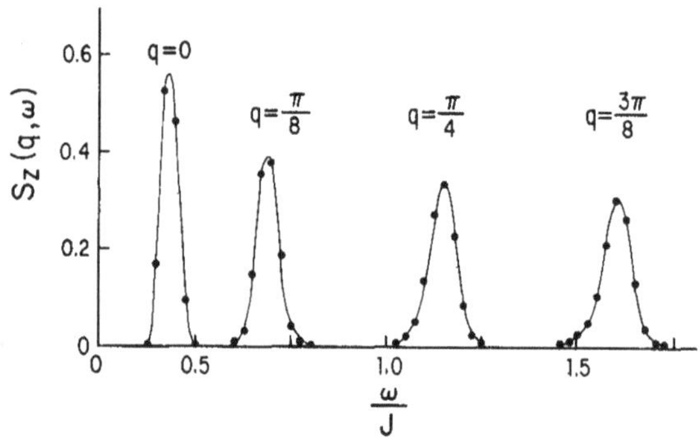

Fig. 2. $S_z(q,\omega)$ as a function of ω for the ferromagnetic XY-chain with $kT = 0.1J$ and $h = 0.1J$. The solid curves are best fits to Lorentzian line shapes convoluted with the Gaussian resolution function.

broadens. The combination[6] of a Gaussian central peak and resolution broadened Lorentzian peak fits the data quite well.

$S_x(q,\omega)$ is expected to be more complicated since both solitons and two-spin-wave processes are expected to play a role. As shown in Fig. 5, at low temperatures a step-like two-spin-wave difference process central peak can be seen at low frequencies, and at a much higher frequency a second peak due to two-spin-wave sum processes appears.

As the temperature is raised the intensities of both peaks increase and the lineshapes begin to change. We believe that the growth of a Gaussian-like central peak reflects the addition of a soliton contribution to the already existent two-spin-wave peak. The temperature dependencies of the half-widths of the central peaks in $S_x(q,\omega)$ and $S_y(q,\omega)$ thus are quite different. The half-width for $S_y(q,\omega)$ increases monotonically with temperature in a manner which is qualitatively similar to the predictions of the sine-Gordon mapping (see Fig. 6). For $S_x(q,\omega)$, however, the halfwidth has a non-monotonic temperature dependence reflecting the change in line-shape which occurs as solitons become thermally excited. In Fig. 7 we compare the behavior of the single spin-wave peak with the theoretical predictions. The intensity of the peak in the y-polarization is adequately described by the sine-Gordon mapping only at the lowest temperatures and the half-width never agrees with theory. The theory

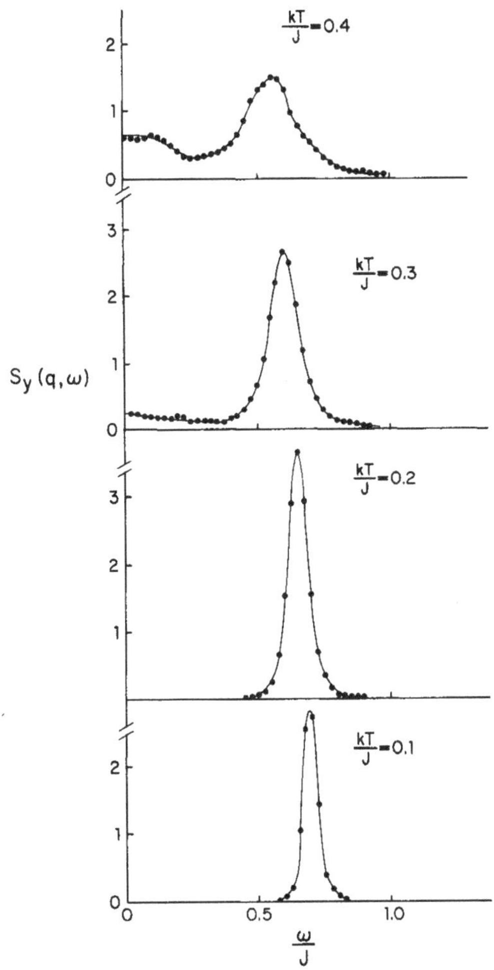

Fig. 3. $S_y(q,\omega)$ as a function of ω for the ferromagnetic XY-chain with $q = \frac{\pi}{8}$ and $h = 0.1J$. The solid curves are fitted intensities representing the sum of resolution broadened Gaussian central peaks and Lorentzian spin-wave peaks.

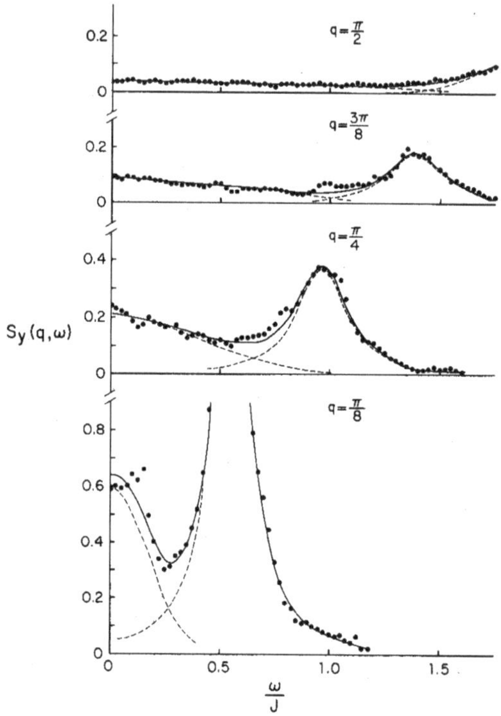

Fig. 4. $S_y(q,\omega)$ as a function of ω for the ferromagnetic XY-chain at $kT = 0.4J$ and $h = 0.1J$. Dashed curves represent individual resolution broadened central peak and spin-wave contributions and the solid curves are the sum of the two.

also restricts the spins to the XY-plane and therefore does not allow any excitations in $S_z(q,\omega)$. In contrast, our data for $S_z(q,\omega)$ show well defined spin-wave peaks for which there is no theoretical description.

4.2 Antiferromagnetic Case ($J < 0$)

The data for the antiferromagnetic chain are much more difficult to interpret than are those for the ferromagnet. The results for $S_\alpha(q,\omega)$ show noticeable differences from run to run and the overall intensities tend to be low. Since there is no in-plane anisotropy, the spins enter a spin-flop state as soon as a non-zero magnetic field is applied. Thus the x-spin component shows a non-zero uniform magnetization whereas the y-spin component shows a non-zero staggered magnetization.

In Fig. 8 we show $S_z(q,\omega)$ for $kT = 0.1J$ and $h = 1.5J$. Instead of the narrow single spin wave peaks which were found for the ferromagnet (see Fig. 2), we observe a rather weak double peak structure in which the relative intensities of the peaks are very q-dependent. (Some of the data for still larger q-values show even more pronounced double peaks.) The dispersion relation which one extracts from these data is thus

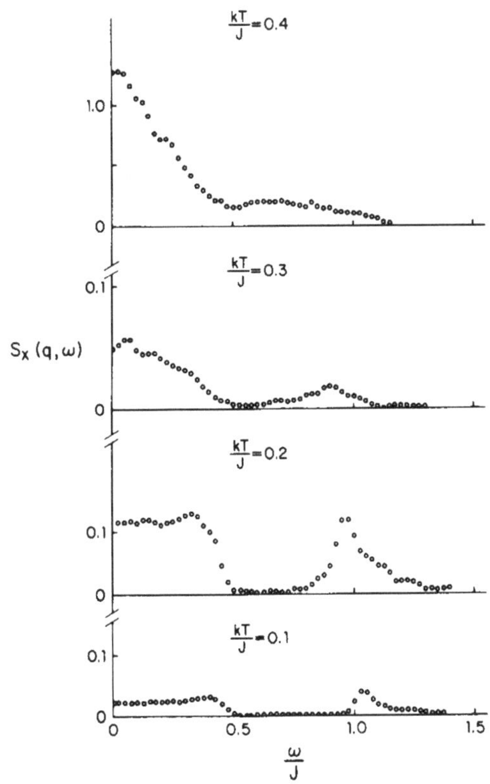

Fig. 5. $S_x(q,\omega)$ as a function of ω for the ferromagnetic XY-chain for $q = \frac{\pi}{8}$, $h = 0.1J$.

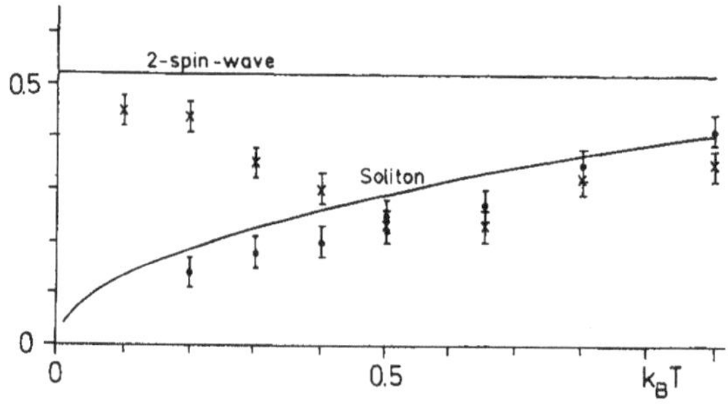

Fig. 6. Half width at half maximum of the central peak in $S_x(q,\omega)-\times$, and $S_y(q,\omega)-\bullet$, as a function of temperature for $h = 0.1J$, and $q = \frac{\pi}{8}$. The solid lines are theoretical predictions (ref. 6).

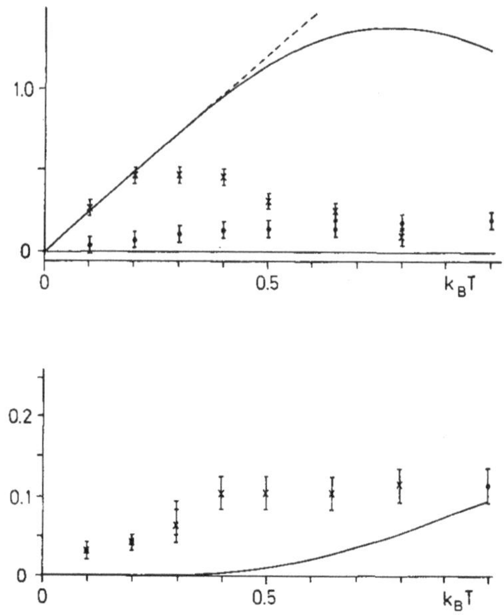

Fig. 7. Temperature dependence of the integrated intensity (top) and half-width at half maximum (bottom) for the ferromagnetic XY-chain with $h = 0.1J$, and $q = \frac{\pi}{8}$. Data are $S_y(q,\omega) - \times$, $S_z(q,\omega) - \bullet$. The dashed line is the leading term from theory and the solid curves include corrections (see Ref. 6).

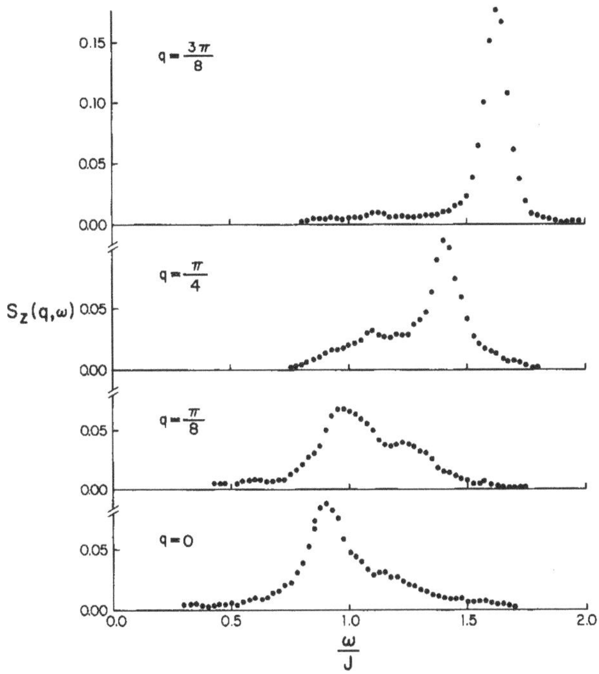

Fig. 8. $S_z(q,\omega)$ vs ω for the antiferromagnetic XY-chain with $kT = 0.1J$ and $h = 1.5J$.

complicated. These data are similar to those observed by Boucher et al.[16] in TMMC and attributed to a new type of two-spin-wave mode which exists along with the single spin-wave peak. Similar structure appears in our data for $S_y(q,\omega)$, but $S_x(q,\omega)$ shows an inverted dispersion curve typical of an aligned paramagnet in a field.

Our data also show central peaks for $q = \pi$ (and for small values of $q^* = \pi - q$ as well) in both $S_x(q,\omega)$ and $S_y(q,\omega)$. The central peaks are quite strong for $S_y(q,\omega)$, as shown in Fig. 9, and can be well fitted by a modified Lorentzian[10]

$$S(q,\omega) \sim \frac{1}{(\omega^2 + \Gamma^2)^{\frac{3}{2}}}. \tag{5}$$

The half-width also increases rapidly with q^* in disagreement sine-Gordon theory[9] but in agreement with the prediction by Maki[10]. The lack of central peaks for $q \sim 0$ indicates that the "π-like" solitons dominate. Data over a much wider range of fields are needed in order to separate the effects of A-type and B-type "π-like" solitons[13,17].

(Note that A-type "π-like" solitons do not exist within the sine-Gordon theory[9].)

5. CONCLUSIONS

High resolution spin dynamics data have been obtained for the XY-ferromag-

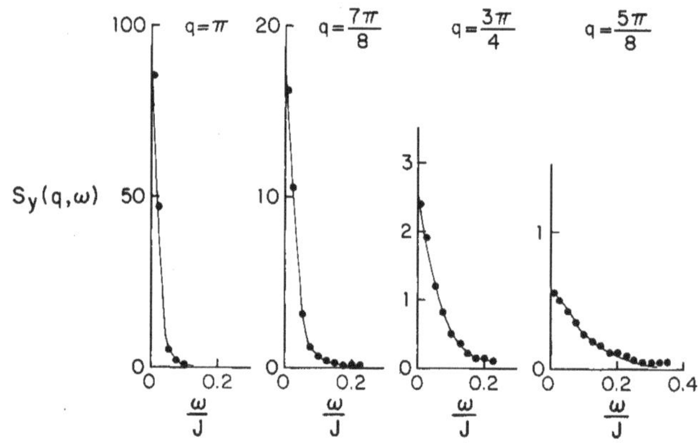

Fig. 9. $S_y(q,\omega)$ vs ω for the antiferromagnetic XY-chain with $kT = 0.3J$ and $h = 1.5J$. The solid curves are fits to Eqn. 5. Note the change in vertical scale for different q-values.

netic chain in a symmetry breaking magnetic field and, to a lesser extent, for the antiferromagnetic chain. For the ferromagnet the predictions of sine-Gordon provide a qualitative description of the behavior but fail quantitatively at all but the very lowest temperatures. Data for the antiferromagnet are more difficult to understand but show evidence of multi- spin-wave behavior and "π-like" solitons. Additional simulations are in progress.

Acknowledgements:

This work was supported in part by NSF grant #DMR-8715740 and ARO grant #DAAL03-87-K-0110. R. W. G. acknowledges the Humboldt Foundation for a Feodor Lynen grant.

6. REFERENCES

1) Steiner, M., Villain, J., and Windsor, C., Adv. Phys. 25, 87 (1976).

2) For reviews see: de Jongh, L. J., J. Appl. Phys. 53, 8018 (1982); Gerling, R. W., Comp. Phys. Rep. 7, 73 (1988).

3) Kjems, J. K. and Steiner, M., Phys. Rev. Lett. 41, 1137 (1978).

4) Steiner, M., Kakurai, K., and Kjems, J. K., Z. Phys. B 53, 117 (1983).

5) Mikeska, H. J., J. Phys. C 11, L29 (1978).

6) Allroth, E. and Mikeska, H. J., Z. Phys. B 43, 209 (1981).

7) Etrich, C., Mikeska, H. J., Magyari, E., Thomas, H., and Weber, R., Z. Phys. B 62, 97 (1985).

8) Boucher, J.P., Regnault, L. P., Rossat-Mignod, J., Renard, J. P. Bouillot, J. and Stirling, W. G., J. Appl. Phys. 52, 1956 (1981).

9) Mikeska, H. J., J. Phys. C13, 2913 (1980).

10) Maki, K., Phys. Rev. B 24, 335 (1981); Maki, K., J. Low Temp. Phys. 41, 327 (1980).

11) Gerling, R. W. and Landau, D. P. in Magnetic Excitations and Fluctuations, eds. S. W. Lovesey, U. Balucani, F. Borsa, and V. Tognetti (Springer, Berlin, Heidelberg, 1984).

12) Staudinger, M., Gerling, R. W., Murty, C. S. S., and Landau, D. P., J. Appl. Phys. 57, 3335 (1985).

13) Gerling, R. W., Staudinger, M., and Landau, D. P., J. Magn. Mat. 54-57, 819 (1986).

14) Landau, D. P., Gerling, R. W., Challa, M. S. S. in Magnetic Excitations and Fluctuations II, eds. U. Balucani, S. W. Lovesey, M. G. Rasetti, and V. Tognetti (Springer, Berlin, Heidelberg, 1987).

15) Gerling, R. W. and Landau, D. P. (to be published).

16) Boucher, J. P., Remoissenet, M., Pynn, R., and Regnault, L. P., J. de Physique 49, C8-1381 (1988).

17) Costa, B. V. and Pires, A. S. T., Sol. State Comm. 66, 1067 (1988).

Early and Late Stages of Spinodal Decomposition

Rashmi C. Desai*

Department of Physics, University of Toronto,

Toronto, Ontario, Canada M5S 1A7

ABSTRACT

I review the phenomenology of the Cahn- Hilliard- Cook (CHC) equation which describes the process of spinodal decomposition. I then outline details of the numerical simulation method that we have used in our recent work on the CHC equation. Simulations were performed to study spinodal decomposition and domain growth following either a critical or an off-critical quench in a two-dimensional system with a continuous, conserved order parameter. In these simulations, the role of thermal noise strength ϵ and the degree of off-criticality ψ_0 is explored. The late stage universal aspects as well as early stage interface formation during the phase separation dynamics are studied. The domain growth behaviour, $R(t) \sim t^{1/3}$ is found to be an asymptotic late time result. The scaled order parameter correlation function becomes independent of ϵ and ψ_o at late times, when interface dynamics play a dominant role.

1. INTRODUCTION

Since the pioneering work of Cahn and Hilliard[1)], there has been a continuing interest in the dynamics of phase separating systems. The nonlinear Cahn-Hilliard-

---

* Research supported by the Natural Sciences and Engineering Research Council of Canada.

Cook (CHC) model[2] is probably the simplest model with a conserved order parameter to study spinodal decomposition and subsequent domain growth. In literature it is also referred to either as model B, as the time- dependent-Ginzburg-Landau model or as the ϕ^4 field theory. The dynamics of first order phase transition has been reviewed in literature including the contributions dealing with spinodal decomposition[3],[4]. In this brief paper, I shall summarise the recent contributions made by us[5],[6],[7],[8].

Spinodal decomposition refers to the phase separation process which occurs in a quenched, thermodynamically unstable system; the eventual equilibrium state for such a process is an inhomogeneous one in which coexisting uniform phases are separated by rather sharp interfaces. The pattern formation during the spinodal decomposition process has been experimentally observed and analysed in alloys, polymer mixtures, glasses and fluid mixtures; it has also been analysed using monte carlo simulations for kinetic ising models and using molecular dynamics for Lennard-Jones fluids. Theoretically the problem of understanding various universal and nonuniversal aspects of such a phase separating system is complicated : it involves (i) thermodynamic instability, (ii) irreversibility, (iii) dynamic creation of interfaces from an initially homogeneous system and (iv) growth of ordered domains for which the boundaries move in time. Initially the thermodynamic instability leads to a rapid growth (exponential in a linear approximation) of long wavelength fluctuations in the order parameter. The nonlinearity in the problem tempers this growth and is responsible for saturation in the value of the order parameter as well as the creation of sharp interfaces between the domains of the separating phases. The late stages of the phase separation process are dominated by the interfacial dynamics. Throughout the nonequilibrium process, the conservation constraint on the order parameter plays a crucial role.

2. THE CHC MODEL

The phenomenology of the CHC model begins with the conservation law for the order parameter ϕ :

$$\frac{\partial \phi}{\partial t} = -\vec{\nabla}.\vec{J}. \tag{1}$$

The local current $\vec{J}$ is related to the gradient of the local chemical potential μ via a generalised constitutive relation in which the effect of other (microscopic) variables in the system is taken into account by the addition of a random thermal noise term :

$$\vec{J} = -M\vec{\nabla}\mu - \vec{j}_{random}, \tag{2}$$

where M is the mobility coefficient, which in this paper is assumed to be independent of ϕ. The local chemical potential μ is then related to a functional derivative of the local free energy F with respect to $\phi(\vec{r},t)$. The random thermal noise term is essential for initiating the segregation process. If one denotes $\vec{\nabla}.\vec{j}_{random}$ by ζ, then the order parameter ϕ satisfies the stochastic partial differential equation:

$$\frac{\partial\phi}{\partial t} = M\nabla^2\frac{\delta F}{\delta\phi} + \zeta, \tag{3}$$

where ζ satisfies the fluctuation dissipation relationship :

$$\langle\zeta(\vec{r},t)\zeta(\vec{r}',t')\rangle = -2k_BTM\nabla^2\delta(\vec{r}-\vec{r}')\delta(t-t'), \tag{4}$$

and the free energy functional $F\{\phi\}$ is given by,

$$F\{\phi\} = \int d^dr[f(\phi) + \frac{\kappa}{2}(\vec{\nabla}\phi)^2]. \tag{5}$$

In the CHC model the bulk free energy term, $f(\phi)$, has a double well structure :

$$f(\phi) = -\frac{r}{2}\phi^2 + \frac{u}{4}\phi^4. \tag{6a}$$

In contrast, in the single phase region $f(\phi)$ has a stable single-well structure and can be approximated as,

$$f_1(\phi) = \frac{r'}{2}\phi^2. \tag{6b}$$

In this description, κ, u, r, r' and M are positive phenomenological (in principle temperature dependent) coefficients. The free energy functional F which is employed in the CHC model (eq.(5,6)) is often referred to as the Ginzburg- Landau-Wilson free energy in the current literature. The anharmonic terms in $f(\phi)$ make eq.(3) nonlinear in ϕ. Substituting the free energy functional, eq.(6a) into eq.(3) reveals the nonlinear Cahn- Hilliard- Cook model :

$$\frac{\partial\phi}{\partial t} = -M\nabla^2\left(\left(\kappa\nabla^2 + r\right)\phi - u\phi^3\right) + \zeta. \tag{7}$$

Typically the phase separation dynamics are monitored through measurement of the time dependent structure factor $S(k,t)$. It is the Fourier transform of the pair correlation function and is defined as,

$$S(k,t) = \frac{1}{V}\int d\vec{r}e^{i\vec{k}.\vec{r}}g(r,t) =< \delta\phi_k\delta\phi_k^* > . \tag{8}$$

The pair correlation function, $g(r,t)$, is defined as,

$$g(r,t) = \frac{1}{V}\int d\vec{r'} < \delta\phi(\vec{r}+\vec{r'},t)\delta\phi(\vec{r'},t) > . \tag{9}$$

$\delta\phi$ represents fluctuations of the order parameter about its average value ϕ_o and $\delta\phi_k$ is its kth Fourier component. Two averages are implicit in eq. (9): the angular brackets, $< ... >$, imply an average over a large number of initial states (i.e., many 'trials' in a numerical simulation) and the integration over $d\vec{r'}$ implies a spatial average. $S(k,t)$ is of considerable theoretical interest because it is an experimentally accessible quantity describing many aspects of phase separation. In particular, the Fourier mode, k_{max}, at which $S(k,t)$ peaks is related to the average domain size and the behaviour of $S(k,t)$ at large wavevectors is related to the structure of the interface separating the domains. At late times when the morphology scales in time, the structure factor S (or the pair correlation function g) for a d-dimensional system can be written in a scaling form, using the single dominant length - the mean domain size $R(t)$ which at late times is expected to grow in a power-law manner $\sim t^x$. The numerical value of the exponent x and the scaling functions S_o and g_o are among the important universal properties characterising the phase separation dynamics of many real systems which have been modelled by the CHC equation (7).

Much of the theoretical framework for understanding the late stage dynamics of phase separation has grown from the classic work of Lifshitz and Slyosov[9)] and of Wagner[9)] who examined the cluster growth for *quenches near the coexistence curve*. Their starting point was the assumption that spherical domains (clusters) exist, and further that the interfaces between the two phases can be approximated by time dependent boundary condition for the diffusion of matter. They developed a description of coarsening based upon an evaporation- condensation mechanism, whereby large clusters grow at the expense of smaller ones. The growth is mediated by the surface tension. Their mean field asymptotic analysis predicts a power law growth for the average cluster radius $\sim R(t)$ with an exponent $x = 1/3$. In the

CHC equation, the surface tension enters through κ, the coefficient of the square gradient term in the free energy functional, eq. (5).

3. SIMULATIONS

Both for the purposes of numerical simulations and later theoretical analysis, it is appropriate to convert eq. (7) to a dimensionless form by making the following transformation of variables[10],

$$\vec{x} = \left(\frac{r}{\kappa}\right)^{1/2} \vec{r},$$

$$\tau = \left(\frac{2Mr^2}{\kappa}\right) t,$$

and

$$\psi = \left(\frac{u}{r}\right)^{1/2} (\phi - \phi_o).$$

Since ϕ is conserved, with its spatial average ϕ_o a time invariant quantity, the dimensionless order parameter fluctuation ψ has its space integral constrained to be zero at all times. In the two phase region the dimensionless equation of motion for ψ is,

$$\frac{\partial\psi(\vec{x},\tau)}{\partial\tau} = -\frac{\nabla^2}{2}\left\{[q_c^2 + \nabla^2]\psi(\vec{x},\tau) - 3\psi_o\psi^2(\vec{x},\tau) - \psi^3(\vec{x},\tau)\right\} + \sqrt{\epsilon}\mu(\vec{x},\tau), \quad (10)$$

where

$$q_c^2 = 1 - 3\psi_o^2,$$

$$\psi_o = \left[\frac{u}{r}\right]^{1/2} \phi_o,$$

$$\epsilon = \frac{k_B T u}{r^2}\left[\frac{r}{\kappa}\right]^{d/2},$$

and

$$\langle\mu(\vec{x},\tau)\mu(\vec{x}',\tau')\rangle = -\nabla^2\delta(\vec{x} - \vec{x}')\delta(\tau - \tau'). \quad (10a)$$

The transformation reveals two dimensionless parameters, ϵ and ψ_o, in the CHC model. $\psi_o = 0$ corresponds to a critical quench (i.e., equal amount of both phases) and $\psi_o = \pm 1/\sqrt{3}$ corresponds to the classical spinodal. All our simulations are done following quenches within the classical spinodal, i.e. with a thermodynamically unstable initial state.

Grant *et al.*[10] have shown that near the critical point (assuming hyperscaling) $\epsilon \sim \xi_o^{-d}$, where ξ_o is the amplitude of the correlation length. This result indicates that even though ϵ may become large, it remains finite at the critical point and is inversely proportional to the interaction length. In simulations performed by us[5-8] ϵ is kept small. This corresponds to either a very deep quench or to systems with large interaction lengths such as polymer systems.

In eq.(10), the overall Laplacian on the right hand side arises from the conservation law; it acts on three terms : the first term is linear in ψ, and leads to a linear instability and an exponential growth of ψ in the initial stages of spinodal decomposition. The second bilinear term treats the two phases in an asymmetric manner and is responsible for nucleation-like droplet growth during the dynamics following an off-critical quench ($\psi_o \neq 0$). This bilinear term is absent for critical quench ($\psi_o = 0$). The third cubic term treats the two phases symmetrically and is responsible for slowing the exponential growth arising from the linear term. It leads to saturation of $\psi(\vec{x}, \tau)$ at its bulk equilibrium values ± 1 and the formation of interfaces. Finally the thermal noise μ satisfies the fluctuation-dissipation relation eq. (10a), which also has an overall Laplacian on its right hand side as a consequence of the conservation law.

At a linear level, the noise averaged eq.(10) leads to an exponential growth of all Fourier components with $k < q_c$. The k-dependent growth rate is $\gamma(k) = \frac{1}{2}k^2(q_c^2 - k^2)$. Even though modes with $k > q_c$ are damped, they play a significant role in the formation of interfaces[11] in the early stages of evolution.

Recently the use of numerical simulations based on a coarse grained description of the order parameter to study the phase separation process has become more extensive. Simulations represent a very powerful method for obtaining insight into the influence of nonlinearities on the dynamics. The simulations described in references [5,6,7,8] represent a two dimensional realization of eq.(10).

3.1. CRITICAL QUENCHES

In references [5,6], critical quenches are studied and the effects of varying (a)the initial state, (b) the noise strength ϵ during the dynamics, and (c)the space and time steps in the discretisation, are explored. For the critical quenches, the algorithm which discretises eq.(10) for the purpose of numerical simulation, uses Euler's finite difference approximation for both the spatial and temporal derivatives. The resulting discretized equation ($\psi_o = 0$ for critical quenches) is

$$\psi_{i,j}(n) = \psi_{i,j}(n-1) + \frac{\Delta\tau}{2\Delta x^2}\sum_{nn}\left[-\psi_{i,j}(n-1) + \psi_{i,j}^3(n-1) - \frac{1}{\Delta x^2}\sum_{nn}\psi_{i,j}(n-1)\right]$$

$$+\mu_{i,j}(n-1). \tag{11}$$

The order parameter ψ is defined only on the sites $i, j \in [1, N]$ of a square lattice. The summation over nearest neighbours is such that

$$\sum_{nn} f(\psi_{i,j}) = f(\psi_{i+1,j}) + f(\psi_{i-1,j}) + f(\psi_{i,j+1}) + f(\psi_{i,j-1}) - 4f(\psi_{i,j}), \tag{12}$$

for an arbitrary function $f(\psi)$. The value of $\{\psi_{i,j}\}$ at a given iteration, n, depends upon $\{\psi_{i,j}\}$ at the previous iteration as well as the thermal noise $\{\mu_{i,j}\}$. Unlike Ising simulations, here ψ is a continuous variable, with equilibrium values of ± 1 corresponding to the wells of the free energy functional defined in eq.(5,6a).

The fluctuation dissipation relation for the discrete equation can be maintained by producing two independent gaussian random variables $\nu^{(1)}$, $\nu^{(2)}$ at each lattice site, such that

$$\mu_{i,j} = \frac{1}{\Delta x}(\nu_{i+1,j}^{(1)} - \nu_{i,j}^{(1)} + \nu_{i,j+1}^{(2)} - \nu_{i,j}^{(2)}) \tag{13}$$

where,

$$< \nu_{i,j}^{(a)}(n)\nu_{k,l}^{(b)}(m) >= \frac{\epsilon\Delta\tau}{(\Delta x)^d}\delta_{i,k}\delta_{j,l}\delta_{m,n}\delta_{a,b}.$$

The mapping manifestly obeys the conservation law appropriate for a critical quench,

$$\sum_{ij}\psi_{ij} = 0. \tag{14}$$

The discretization scheme introduces a (dimensionless) timestep, $\Delta\tau$, and a (dimensionless) mesh size, Δx. Connection with the continuous equations (eq.(10,10a)) can be established by noting,

$$\tau = n\Delta\tau$$

and

$$\vec{x} = i\Delta x\hat{x} + j\Delta x\hat{y}.$$

The dynamical evolution of the discrete and continuous models, however, are not necessarily identical. For example, the discrete equations can undergo a subharmonic bifurcation for a certain range of Δx and $\Delta\tau$; whereas, this is not possible in the continuous equations. As a result, the timestep and mesh size must be chosen carefully in order to establish a correspondence between the two models. These restrictions are discussed in reference [5].

The initial state of the system is usually prepared in a homogeneous (single phase) state by assigning, to each lattice site, a random number uniformly distributed in the interval $[-\epsilon_o, +\epsilon_o]$. The magnitude of ϵ_o reflects the strength of thermal fluctuations in the initial state. All computations were carried out on a 2-d square lattice with periodic boundary conditions. Tables I in reference [5] summarize the parameters used for each run.

3.2. OFF-CRITICAL QUENCHES

In the study of off-critical quenches[7),8)], we only considered simulations of the deterministic Cahn-Hilliard model which is eq.(10) with $\epsilon = 0$. The examination of late stage phenomena in particular, is always constrained by computer resources, so it was necessary to maximize the efficiency of the numerical approach. In the absence of thermal noise ($\epsilon = 0$), eq.(10) becomes deterministic and one has a greater flexibility in the choice of algorithm.

The Euler discretization, used for the stochastic simulations above, is limited by a numerical instability as manifested in the subharmonic bifurcation condition[5)]. For small Δx, this condition has the structure

$$16\frac{\Delta\tau}{\Delta x^4} < 1. \tag{15}$$

Large timesteps, which violate the inequality, destabilize the large k modes causing an oscillating checkerboard pattern. (The simulations quickly become divergent after this). The instability is rooted in the intrinsic structure of the field theory, where small lengthscale fluctuations (i.e. high k-modes) are stabilized by the highest order Laplacian. Thus in eq.(10), fluctuations in these modes are dominated by a diffusion-like equation

$$\frac{\partial\psi}{\partial\tau} = -\frac{1}{2}\nabla^4\psi. \tag{16}$$

The restriction on $\Delta\tau$, as expressed by eq.(15) can be interpreted physically as the requirement that the timestep be smaller than the 'diffusion time' across a spatial cell of dimension Δx, (apart from a constant factor). This represents an intrinsic timescale for the decay of local fluctuations.

More generally, the subharmonic bifurcation condition is an indication of an inherent stiffness in the formulation of the domain growth problem. The stiffness emerges because there are two timescales in the system; namely, the rapid decay of localized (small wavelength) fluctuations, and the slow growth of the overall domain morphology. Although the latter is the phenomenon of interest , the more rapid timescale sets the limitations on the computation.

The numerical algorithm can be improved by using implicit time discretization. Such methods involve an evaluation of the right hand side of eq.(10) at the nth timestep (as well as previous timesteps). The result is an implicit set of coupled nonlinear algebraic equations for $\psi_{ij}(n)$ which must be solved at each timestep. This approach has the advantage that the implicit equations are usually very stable. However, the drawback is that a set of nonlinear equations must be solved for each iteration. Typically, the equations are solved using a predictor-corrector approach. (For stiff equations, however, the predictor-corrector method often possesses instabilities).

For the off-critical quench simulations, a fourth-order Adams-Bashforth predictor and Adams-Moulton corrector were adopted because of their stability and efficiency. A timestep of 0.1 was used, which led to an average fractional correction to ψ of approximately 10^{-6} at each iteration during the time regime of interest. Early stage phenomena are more sensitive to the timestep where a timestep of 0.01 was used and the late stages of some runs were repeated with a timestep of 0.01 to insure convergence. (It should be noted that the Adams-Bashforth Adams-Moulton algorithm also becomes unstable for large timesteps as a result of the stiffness in the problem).

Spatial gradients in eq.(10) were approximated by an 'isotropic' form for the discrete Laplacian which couples each cell to nearest and next-nearest neighbours. For an arbitrary function $f(\psi)$,

$$\begin{aligned}\nabla^2 f(\psi) \rightarrow & \frac{1}{4\Delta x^2}\left[f(\psi_{i+1,j+1}) + f(\psi_{i-1,j+1} + f(\psi_{i+1,j-1}) + f(\psi_{i-1,j-1})\right] \\ & + \frac{1}{2\Delta x^2}\left[f(\psi_{i+1,j}) + f(\psi_{i-1,j} + f(\psi_{i,j+1}) + f(\psi_{i,j-1})\right] \\ & - \frac{3}{\Delta x^2} f(\psi_{i,j}). \end{aligned} \tag{17}$$

(This discretization should be more stable than the nearest neighbour coupling given in eq.(12). However, we found that it made no difference for the quantities that were measured). The meshsize Δx introduced in the spatial discretization is an important consideration for late stage phenomena. It is advantageous to make the meshsize large since the time required for a simulation varies approximately as Δx^{-4}, (see eq.(15)). However, Δx is bounded by the sharp gradients inherent to the interface. (Again, this is a limitation intrinsic to the formulation of the domain growth problem). It was found that the morphology of the clusters also seems to play an important role, with circular clusters being more sensitive to Δx than the convoluted domains seen in critical quenches. Large meshsizes typically generate anisotropic clusters which align with the underlying lattice imposed in the difference scheme. In addition, cluster growth is slowed down. For the simulations reported here the meshsize was set at 1.0.

Most of the late stage simulations were performed on square lattices of size 256^2 using periodic boundary conditions. However, smaller lattices (128^2) were also examined to insure that there were no finite size effects. Most of the early stage results were obtained using 100^2 lattices.

4. SIMULATION RESULTS

The results of the simulations have been described in detail elsewhere[5)-8)]. Here I give summary highlights.

CRITICAL QUENCHES. Early time simulations following a critical quench[6)] into the thermodynamically unstable part of the phase diagram helped to delineate the regions of validity of linear and some early time nonlinear theories of spinodal decomposition. And these simulations also confirmed a suggestion by Binder that the time of validity τ_{max} of the linear theory is proportional to $ln(\epsilon)$. Surprisingly, other early time nonlinear theories also had similar logarithmic behaviour with somewhat larger numerical coefficients.

The pattern generated during the phase separation following a critical quench is invariably an interconnected, interwoven spatial distribution of positive and negative values of the order parameter, since the instability treats the two phases symmetrically. (See figure 1 in reference [6] and figures 1 and 6 in reference [5].) At late times the patterns become self-similar. At early times the dominant wavenumber describing the morphology is comparable to ξ^{-1}, where ξ is the late stage interfacial width. Subsequent to the initial linear regime, the interfaces separating the domains begin to sharpen. Asymptotically, the morphology would be characterised by domains (where ψ is near one of the two equilibrium values ± 1), separated by relatively sharp interfaces. The approach to the asymptotic configuration depends on the strength of the thermal noise. For moderate values of ϵ, the interfaces appear rough indicating diffuse boundaries; larger the value of ϵ, longer it takes to create sharp boundaries.

At late stages there are two important lengths: the mean domain size $R(t)$ which is growing as a power law in time and the time independent interface width ξ. Figure 1 shows[12)] the relationship between ξ and $R(t)$ for a late stage configuration. Here the interface is defined as a region over which the order parameter varies from $A\psi_+$ to $A\psi_-$, where $\psi_\pm$ are the final equilibrium values of the order parameter and A is arbitrarily set at 0.75. The upper plot (Fig.1a) demarks the interfacial sites of the system and the lower plot gives a crosssectional view of the variation of ψ as the system is traversed along the horizontal line shown in Fig.1a. The sharpness of the interface is clearly evidenced in the steep gradients in Fig.1b. The late stage behaviour is intimately connected with the emergence of sharp interfaces.

The quantitative analysis of domain growth is described in references [5,6] and shows that the symmetric (critical) quench gives a late stage growth exponent $x = \frac{1}{3}$. It also shows that the scaling of the pair correlation function $g(r,t) = (1 - \psi_o^2)^{-1} g_o(r/R(t))$, is asymptotic where thermal fluctuations become irrelevant. (see figures 2,3,4,5 and 8 in reference [5].) The best data for the scaled pair correlation function $g_o(z)$ at the latest time from our simulations are shown in Figure 9 of reference [7]. The figure includes $g_o(z)$ for $\psi_o = 0$ (critical quench) and $\psi_o = 0.4$. It appears very likely that the asymptotic scaled pair correlation function (as defined above) is independent of both the parameters, ϵ and ψ_o, in the CHC model.

OFF – CRITICAL QUENCHES. For nonzero ψ_o, the spatial symmetry between the two phases is broken. In order to study the consequences of this, we performed the off-critical quench study[7),8)], where the thermal noise strength ϵ was set to zero and ψ_o was varied. The ψ_o-scaling of $S(k,t)$, implied by the linear approximation to the CHC model, is described in detail in reference [8] with the results shown in Figure 4 there. The wavenumber k scales to $\frac{k}{q_c}$ and the time τ scales to τq_c^4. Such a scaling could provide a useful procedure to analyse the early stages of the off-critical quench experiments. The simulations show that even at very early times the large k-modes do not obey such a scaling. The nonlinearity which is essential in the interface formation shows up in the large k-modes at the earliest of times.

During the early stages of interface formation in the off-critical quench simulations, the bilinear and cubic terms in eq.(10) compete with one another: The

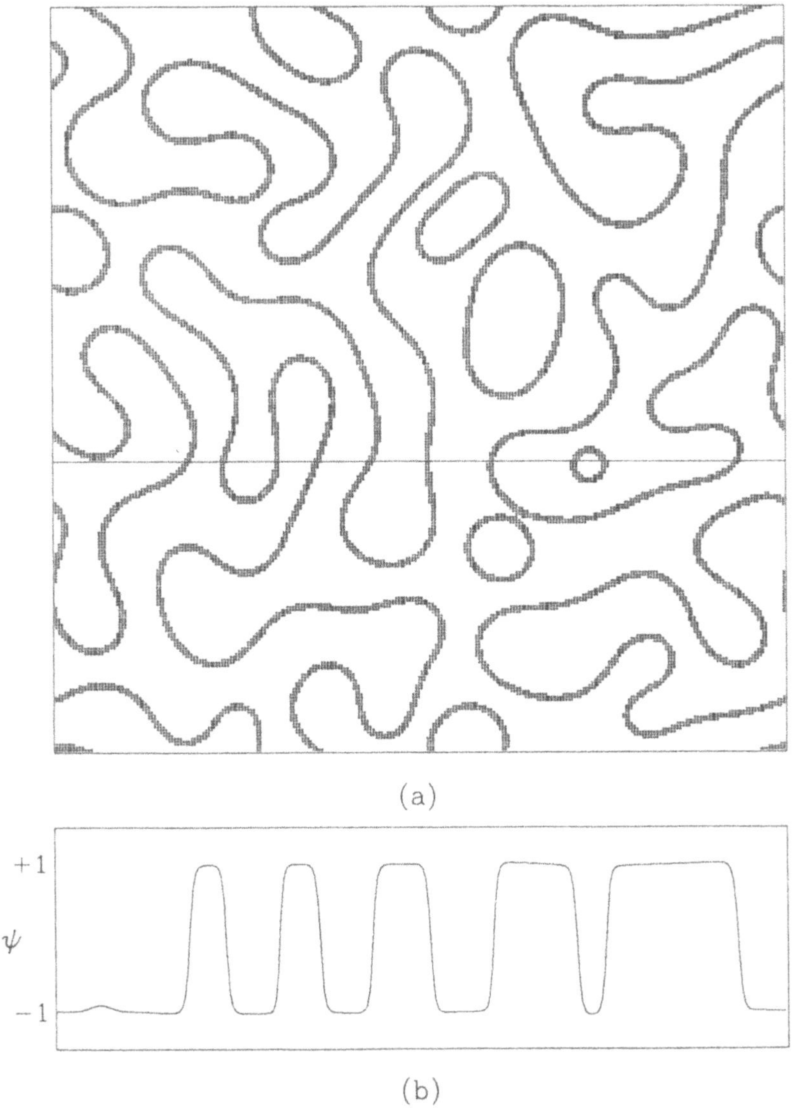

Figure 1 Interfacial structure for late stage configurations ($\epsilon = 0$). In (a) the shaded regions correspond to interfaces separating the domains as defined in the text. A cross-sectional view of the order parameter is given in (b). The location of the cross-section is denoted by the line in (a).

asymmetric (*bilinear*) term mainly couples the Fourier component ψ_k to all its even harmonics ψ_{2k}, ψ_{4k}, etc. As a consequence the growth rate in the bulk of the minority phase is increased while that in the bulk of the majority phase reduced; at the boundaries of the majority and minority domains, the behaviour is reversed and one sees an increased growth rate of the majority phase and a decreasing growth of the minority phase. The net effect is to sharpen domain walls within the background and to decrease the spatial extent of the minority phase, with a consequence that the mean domain size $R(t)$ decreases with t in the early stages. (See Figure 5 in reference [7].) This feature is absent in the critical quench simulations, since the *symmetric* (*cubic*) term treats the majority and minority phases without any bias; it couples only odd harmonics and slows the growth of both phases. The symmetric term opposes the growth in the bulk region and favours it at the boundaries. Thus while the asymmetric term decreases the spatial extent of the minority phase, the symmetric term is mostly responsible for sharp interfaces and subsequent coarsening of domains. Figures 7 and 8 of reference [8] show these effects in the morphology and two-point distribution function respectively. For $\psi_o = 0$, one typically has convoluted morphology; this is also the case at early time for $\psi_o > 0$. For this later case however there is a crossover to a morphology dominated by circular clusters. Smaller the value of ψ_o, longer the crossover takes; this observation is expanded in Figures 5,6 and 7 in reference [8].

The late stage coarsening for the off-critical quenches is described in detail in reference [7]. From the early stage study, it is clear that the smoothening of interfaces to form circular domains play an important role and has the effect of extending the timescale for the onset of scaling regime where the morphology is one of circular domains if $\psi_o > 0$. The role that the asymmetry in phases plays in the growth law and in scaling is explored in reference [7] by monitoring the time evolution of various measures of domain size. These are (i)the mean domain size as obtained from $g(r)$, (ii)the inverse perimeter density of the domains of the minority phase, (iii)the average number of clusters per unit area and (iv)the average cluster radius as measured by the radius of gyration. For the percolating-domain morphology resulting after a critical quench, only the first two can be defined. The measures (iii) and (iv) are related to the cluster size distribution. The onset of a scaling regime has to be characterised by the establishment of scaling behaviour in each of such measures that one can devise. (See table II in reference [7].) From the simulations with nonzero ψ_o, it is found that smaller the value of ψ_o, longer it takes to establish the scaling behaviour and the circular domain morphology.

For quenches near the coexistence curve, the classic analysis of Lifshitz and Slyosov[9] (LS) shows that clusters of minority phase compete through an evaporation condensation mechanism, whereby larger clusters grow at the expense of smaller ones and the cluster interaction is mediated by diffusion in the majority phase. In the late stage study the quench closest to the coexistence curve is with $\psi_o = 0.4$. Figure 2 shows[12] the evolution of the cluster boundaries for a typical realisation. The growth of large clusters at the expense of smaller ones is clearly evident. Cluster coalescences do not seem to play a major role despite the fact that the minority phase is quite concentrated. The figure demonstrates however that the dynamics of the clusters depends on their immediate environments. An example of this is seen in the evolution of clusters labelled A and B in the figure. For the earlier times shown in the figure, the two clusters are roughly the same size. Nonetheless, cluster A, which is surrounded by smaller clusters grows in time, whereas cluster B, which is surrounded by larger clusters shrinks to zero.This type of cluster correlations are absent in the LS analysis which has in it an inherent dynamic mean field approximation. Understanding such correlations and finding an analytic representation for the asymptotic cluster size distribution (see appendix in reference [7]) in the domain growth problem is currently an important unfinished task.

The role of ψ_o in the late stages of coarsening is rather subtle. The value of the field at which the domains saturate ($\psi_\pm = \pm 1$), and the late stage interfacial profile do not depend on this parameter. As a result, the dynamical equations describing the motion of the interfaces are independent of ψ_o. However, the position of the quench does affect the amount of each phase present in the late stages, as well as the spatial distribution of these phases. In this manner, there is an indirect influence on the growth properties.

In the context of domain coarsening, the role of early stage instabilities is to generate the initial topological structure of the domains. The convoluted pattern often associated with spinodally decomposing systems is a result of the linear instability. However, even within the mean field spinodal of the CHC model, nonlinear terms act to break the symmetry and can generate circular domains which are more often associated with nucleation. The transition from convoluted domains to circular clusters is smooth. The classical spinodal (which is a prediction

Figure 2 Evolution of clusters for $\psi_o = 0.4$. This plot shows the interfaces separating the clusters ($\psi < 0$) from the matrix ($\psi > 0$) during the late stages of evolution. In the figure, the interfaces for two different times are superimposed in order to track the evolution of individual clusters. The lighter lines correspond to $\tau = 1000$; and the darker lines, to $\tau = 5000$. The size of the cluster A is seen to increase during this time interval, while the cluster B shrinks and disappears.

of linear theory) is of little consequence for late stage phenomena in this model, since nonlinear instabilities play an increasingly more important role as the system is moved off-critical. Our simulation results indicate that such a crossover to a circular- cluster- morphology should occur even for the smallest *nonzero* ψ_o as soon as the spatial symmetry between the two phases is broken.

In the late stages of phase separation, surface tension is the driving force for domain coarsening. For the quench to $\psi_o = 0.4$, the domains of the minority phase remain circular as coarsening progresses. The growth proceeds by diffusive transport from the smaller clusters to the larger ones. The time evolution of the first zero of the pair correlation function, the average cluster radius, the interfacial density and the cluster number density are consistent[7)] with a scaling morphology. Moreover, the scaled pair correlation function and the scaled cluster probability density exhibit very little change with time during the late stages. (However, the statistics for the latter scaling function are rather poor). These results are a very strong indication that there is dynamical scaling. The effective exponent describing the evolution of the pattern is less than 1/3, which most likely indicates that the simulations have not reached the asymptotic regime for this quench.

For smaller values of ψ_o, domain interaction is more intense in the early stages, resulting in convoluted, noncircular domains. Here, smoothening of the surface plays an important role. This coarsening mechanism is precluded by the assumptions of LS. Quenches through the critical point show that the process leads to a growth exponent of 1/3 for the inverse perimeter density and for the pair correlations. At $\psi_o = 0.1$ and 0.2 there is a crossover from growth dominated by surface smoothening to LS type growth as the system evolves. Late stage scaling of the *morphology* has not set in for these simulations. We conjecture that percolating clusters and circular domains form two types of scaling morphology. The former occurs for critical quenches, while the latter applies near the coexistence curve. When $\epsilon = 0$, the crossover from percolation to isolated droplet growth occurs very close to the critical quench, (i.e. $\psi_o < 0.1$). For all values of ψ_o we have probed, however, our results are consistent with an asymptotic growth exponent of 1/3.

ACKNOWLEDGEMENTS

The work reviewed here is done in collaboration with K.R. Elder[11)] and T.M. Rogers[12)]. It is a pleasure to thank both of them. The late stage simulations were performed on the CRAY XMP at the Ontario Centre for Large Scale Computation; the assistance of the OCLSC-staff is gratefully acknowledged.

REFERENCES

1) Cahn, J.W. and Hilliard, J.E., J. Chem. Phys. 28, 258 (1958); Cahn, J.W., Acta Metall. 9, 795 (1961).

2) Cahn, J.W., Trans. Metall. Soc. AIME 242, 166 (1968); Cook, H.E., Acta Metall. 18, 297 (1970).

3) Gunton, J.D., San Miguel, M. and Sahni, P.S. in *Phase Transition and Critical Phenomena*, Ed. C. Domb and J.L. Lebowitz (Academic, New York, 1983), vol. 8, p. 267.

4) Binder, K., Spinodal Decomposition, ch.7 in *Materials Science and Technology*, vol.5 : *Phase Transformations in Materials*, Ed. P. Hassen (VCH Verlagsgesellschaft Weinheim, Germany, 1990).

5) Rogers, T.M., Elder, K.R. and Desai, R.C., Phys. Rev. B37, 9638 (1988).

6) Elder, K.R., Rogers, T.M. and Desai, R.C., Phys. Rev. B38, 4725 (1988).

7) Rogers, T.M. and Desai, R.C., Phys. Rev. B39, 11956 (1989).

8) Elder, K.R. and Desai, R.C., Phys. Rev. B40, 243 (1989).

9) Lifshitz, I.M. and Slyosov, V.V., J. Phys. Chem. Solid 19, 35 (1961); Wagner, C., Z. Electrochem. 65, 581 (1961). See also Voorhees, P.W., J. Stat. Phys. 38, 231 (1985).

10) Grant, M., San Miguel, M., Vinals, J. and Gunton, J.D., Phys. Rev. B31, 3027 (1985).

11) Elder, K.R., Ph.D. Thesis, University of Toronto (1989), unpublished.

12) Rogers, T.M., Ph.D. Thesis, University of Toronto (1989), unpublished.

ASPECTS OF CLASSICAL EXCITATIONS IN TWO-DIMENSIONAL MAGNETIC SYSTEMS

G.M. Wysin
Department of Physics
Kansas State University
Manhattan, KS 66506 USA

ABSTRACT

Dynamic properties of classical spin excitations in models for quasi-two-dimensional materials are described. Included are spin configurations for vortex and vortex-like excitations from continuum limit equations of motion, together with comments about their stability. Models studied are the anisotropic ferro- and antiferromagnets, and an easy-plane ferromagnet with 4-fold in-plane symmetry breaking. Results from numerical simulations of the dynamic structure function $S^{\alpha\alpha}(\vec{k},\omega)$ are shown and compared with phenomenological theories.

1. INTRODUCTION

1.1 Low-Dimensional Models

Understanding the dynamic properties of excitations in magnetic models is a challenging and exciting enterprise, and from the theoretical viewpoint, analysis for systems in one or two dimensions can in many cases be carried out with more details than would be possible in three dimensions. From the experimentalists' perspective, there exist a wide variety of real materials which can be quite accurately described in terms of strong spin-spin interactions acting only along one or two of the crystal axes, thereby imposing a lower effective dimensionality.[1-4] For these materials, at temperatures above a transition temperature where the full three-dimensional interactions become relevant, the spin degrees of freedom can be described by a Hamiltonian with a lower effective dimensionality $d < 3$. Then the material is called "quasi-d-dimensional."

1.2 Quasi-One-Dimensional Magnets and Solitons

Historically, the most well known quasi-one-dimensional magnetic materials[1] are the spin-1 easy-plane ferromagnet $CsNiF_3$, and the spin-5/2 easy-plane antiferromagnet $(CH_3)_4NMnCl_3$, known as TMMC. These and other materials have been thoroughly studied experimentally and analyzed in terms of nearest neighbor classical spin chain Hamiltonians possessing soliton modes.[5-7] The analysis of equilibrium and dynamic properties for such models is a mature and still active field (for example, see the article by D.P. Landau herein). Usually the interactions are anisotropic, perhaps with an Ising (easy-axis) or planar (easy-plane) symmetry. Presence of Ising anisotropy will lead to the possibility of static soliton-like excitations, involving rotations of the spin field through π, connecting regions having the two energetically equivalent alignments along the Ising axis. With planar anisotropy, there is the possibility of static soliton-like excitations involving in-plane rotations through 2π for ferromagnets or π for antiferromagnets. These typically will be better defined if there is a small applied magnetic field within the easy plane, breaking the symmetry in that plane.

Quite generally, with reasonable continuum limit assumptions, magnetic Hamiltonians in one dimension map approximately onto the one-dimensional sine-Gordon (sG) equation.[5-7] Therefore, phenomenological descriptions involve an "ideal gas" of sG solitons moving within (and perturbing) a background field of small amplitude spin waves.[8] Within this sG description, contributions to the dynamic structure function, $S^{\alpha\alpha}(\vec{k},\omega)$, as measured in inelastic neutron scattering, are expected due to solitons (peaks at $\omega = 0$), spin waves (peaks at finite ω) and multi-spinwave processes (which can also contribute to $\omega = 0$ intensity). In terms of gross features this description is probably valid for many materials. However, there are known difficulties. Corrections are expected due to deviations from sG dynamics.[9,10] The mapping onto the sG equation is approximate---spins tilt out of the easy plane for "planar" systems, an effect not fully accounted for in the sG equation. Also there could be quantum effects, and higher order soliton-spinwave interaction effects. However, if we consider quantum effects to be small (expected for $S \geq 1$), then numerical spin dynamics integrations are a productive way to avoid the sine-Gordon mappings and at the same time obtain accurate results for $S^{\alpha\alpha}(\vec{k},\omega)$. Conveniently, at this time available computing power is appropriate for performing these spin dynamics calculations for reasonably large systems in two dimensions (100 x 100) as well as for one-dimension.

1.3 Quasi-Two-Dimensional Magnets and Vortices

Understanding of the statics and dynamics of quasi-two-dimensional magnets is not so thoroughly developed. However, the variety of challenging models and materials and possibilities for topological excitations is much greater than for one dimension. Some typical examples of quasi-two-dimensional magnets include[2] $BaCo_2(AsO_4)_2$, an XY ferromagnet, $BaNi(PO_4)_2$, an XY antiferromagnet, and Rb_2CrCl_4, an XY ferromagnet with in-plane symmetry-breaking.[3] Also, there is the nearly isotropic Heisenberg ferromagnet K_2CuF_4, and the XY ferromagnet $CoCl_2$ intercalated in graphite.[4] Among the available materials spin-S values range from 1/2 to 2, with interesting exchange and crystal field anisotropies on triangular, honeycomb and square lattices.

For two-dimensional models, in addition to spin waves and solitons (or now, "domain walls"), there can also be "vortex" excitations involving a singular point in the spin field, which has a nonzero curl. All of these can be expected to make contributions to $S^{\alpha\alpha}(\vec{k},\omega)$. The gross features should be well described by a phenomenological model, in terms of an ideal gas of walls, vortices, and spinwaves weakly interacting. For a simple model such as one with planar anisotropy but no applied fields (e.g., XY model), only vortices and spinwaves are present, and an ideal gas phenomenology has been developed by Mertens et al.[11] to describe the dynamics above the vortex unbinding[12] transition temperature (Kosterlitz-Thouless transition temperature T_{KT}). Static vortices are responsible for zero frequency intensity in $S^{\alpha\alpha}(\vec{k},\omega)$, that is, a central peak. The phenomenology can be compared with numerical simulations of $S^{\alpha\alpha}(\vec{k},\omega)$, however, an exact separation of central peak intensity into vortex and multi spinwave components is difficult. The numerical simulations suggest that corrections to the ideal gas phenomenology are needed, especially for the out-of-plane spin component correlations. However, for $T > T_{KT}$, the phenomenology assumes unperturbed vortices moving with a Boltzmann distribution of velocities, but at present it is not completely understood how strongly the vortex-spinwave or vortex-vortex interactions modify the correlations, nor are the dispersion relation and velocity distribution of moving vortices actually known.

Ideally, one should consider the linearized perturbation about a vortex solution to obtain the vortex-spinwave scattering properties and vortex stability with respect to perturbations. In practice, this leads to a 4th order eigenvalue problem for which the bound states are not known. Instead, numerical integrations can be used, starting from a vortex initial condition, to indicate the dynamic stability. Similarly, numerical simulations have been used to indicate properties for travelling vortices. Any information about dynamic properties of isolated vortices obtained in this way will be helpful in improving phenomenological models, and offer guidance for other models involving more complex anisotropic interactions.

There are three primary goals for this paper: i) to review the classical mechanics description of spin waves, domain walls, and especially, vortices in several magnetic models; ii) to present some results relating to stability of vortices (as derived from continuum theory) when placed on a lattice; iii) to present results for $S^{\alpha\alpha}(\vec{k},\omega)$ for these models, in the light of available phenomenology theories. The models considered are the anisotropic Heisenberg model (ferro and antiferromagnetic exchange), and a model for Rb_2CrCl_4, which is an easy-plane ferromagnet with 4-fold in-plane symmetry breaking anisotropy. The intention is to give an idea of the remarkable assortment of excitations in the various models, concentrating on their spin configurations, stability and dispersion properties, and the implications of these for their contributions to dynamic space-time correlation functions.

2. ANISOTROPIC HEISENBERG MODELS: XY TO ISOTROPIC FERROMAGNETS

For spin variables $\vec{S}_i$, where i labels a lattice site in two dimensions, a simple Hamiltonian is

$$H = - J \sum_{(i,j)} (S_i^x S_j^x + S_i^y S_j^y + \lambda\, S_i^z S_j^z) \ , \qquad (1)$$

where the sum is over nearest neighbor pairs (i,j) and the parameter λ measures the degree of exchange anisotropy. Here we consider only planar anisotropy, with $0 \leq \lambda \leq 1$. As λ ranges from 0 to 1 the model changes smoothly from the XY model to the isotropic Heisenberg model.

The discrete equations of motion resulting from (1), obtained most easily through classical Poisson brackets, are

$$\frac{d\vec{S}_i}{dt} = \vec{S}_i \times \vec{F}_i \tag{2}$$

$$\vec{F}_i = J \sum_{(i,j)} (S_j^x \hat{x} + S_j^y \hat{y} + \lambda \ S_j^z \hat{z}). \tag{3}$$

The sum in (3) involves only the nearest neighbors of site i. This form of the differential equations is used for numerical integrations due to its simplicity and lack of trigonometric function evaluations. Using these equations, and assuming small dynamic fluctuations from a state with $\vec{S}=S(1,0,0)$ leads to the spinwave dispersion relation for an excitation with wavevector (k_x,k_y),

$$(\omega/2JS)^2 = (2-c)\ (2-\lambda c); \quad c = \cos k_x a + \cos k_y a \tag{4}$$

where a is the lattice spacing. This dispersion has an interesting feature. For small wavevectors, we have

$$\omega \approx JS(ka)^2 \qquad \text{for} \qquad \lambda = 1 \text{ (Heisenberg)} \tag{5}$$

$$\omega \approx 2JS(ka) \qquad \text{for} \qquad \lambda = 0 \text{ (XY)}. \tag{6}$$

This point is that for intermediate values of λ, there is a mixture of linear and quadratic wavevector dependencies, with a crossover from linear to quadratic k-dependence at adequately large k.

2.1 A Continuum Limit

A crossover effect for the continuum solutions as λ ranges from 0 to 1 might also be expected, especially since a planar vortex is not expected for the isotropic model. The continuum limit[13] can be studied by using spin variables in terms of the spherical coordinates:

$$\vec{S}(\vec{r}) = S(\cos\theta\cos\phi,\ \cos\theta\sin\phi,\ \sin\theta). \tag{7}$$

In a lowest order continuum limit derived from a square lattice, the coupled equations of motion for the $\theta(\vec{r})$ and $\phi(\vec{r})$ fields are

$$\dot{\theta} = JS\ (\cos\theta\ \nabla^2\phi - 2\sin\theta\ \vec{\nabla}\phi\cdot\vec{\nabla}\theta) \tag{8}$$

$$\dot{\phi}\cos\theta = -JS\ \{[|\nabla\theta|^2 + |\nabla\Phi|^2 - 4 + \lambda\ (4-|\nabla\theta|^2)]\ \sin\theta\ \cos\theta + (\sin^2\theta + \lambda\ \cos^2\theta)\ \nabla^2\theta\} \tag{9}$$

These equations have two types of soliton-like vortex solutions[14] (or, an "instanton" when $\lambda = 1$).

2.2 Planar Vortices

Angle ϕ measures the spin angle projected on the easy xy plane, while angle θ measures the tilting of the spins up out of the xy plane. In the XY limit one expects a static planar solution, with $\theta = 0$ and $\dot{\theta} = \dot{\phi} = 0$. This exactly satisfies (9) and then (8) simplifies to the Laplace equation, for any value of λ:

$$\nabla^2\phi = 0\ . \tag{10}$$

This has a simple "planar vortex" (or antivortex) solution

$$\phi = \pm\ \tan^{-1}(y/x) + \phi_o$$
$$\theta = 0 \tag{11}$$

where ϕ_o is an arbitrary constant. The energy depends logarithmically on the system size R and a short distance cut-off r_a,

$$E_{pl} = \pi\ JS^2\ \ln(R/r_a). \tag{12}$$

This is the well known vortex (antivortex) of the XY model, but it also exists for nonzero λ. The vortex is singular at its center; the spin direction is undefined there. The antivortex differs from the vortex only in the sense of the circulation in the spin field.

In a system in thermodynamic equilibrium, it is expected that these are created as vortex-antivortex pairs, where the cost in energy depends logarithmically on the separation of their centers. In the Kosterlitz-Thouless transition for the XY model, it becomes possible to lower the free energy by creating pairs when the creation energy ΔU_{pair} can be counterbalanced by the entropy term $T\Delta S_{pair}$. This occurs at a transition temperature[15] $T_{KT} \approx 0.8\ JS^2$ for $\lambda=0$. It is useful to see whether peculiarities of the classical magnetic Hamiltonian may affect this transition, and consider the dynamic signatures of the unbinding vortices.

2.3 Out-Of-Plane Vortices

It can be seen that equations (8) and (9) admit another static vortex solution,[13,14] possessing a nonzero out-of-plane component θ. With the in-plane angle ϕ given by (11), one can find asymptotic solutions to (9) with the following behaviors for small r and large r:

$$\sin\theta \approx \pm S\ [1 - \tfrac{1}{2}\, a\ (r/r_v)^2] \qquad r \to 0 \tag{13}$$

$$\sin\theta \approx \pm bS\ (r_v/r)^{1/2}\ \exp(-r/r_v) \qquad r \to \infty\ . \tag{14}$$

Here a and b are constants, and r_v is an effective "vortex radius"

$$r_v = \frac{1}{2}\left(\frac{\lambda}{1-\lambda}\right)^{1/2} \ . \tag{15}$$

This vortex is not singular at its center; the spins point along the z-axis at the center, with the z-component dying off exponentially away from the center over a length scale r_v. This configuration is called an "out-of-plane vortex."

Although an exact analytic expression for the out-of-plane vortex is not known, Gouvea et al.[14] have made an approximate evaluation of its energy E_{out}. The result is that for $\lambda << 0.8$, the planar vortex is energetically favorable; $E_{pl} < E_{out}$. But for $\lambda > 0.8$, the relationship is reversed; $E_{pl} > E_{out}$. This is an intriguing result; for a given value of λ, are both types of vortices possible, or is only the energetically favorable type dynamically allowed?

2.4 Vortex Dynamic Stability

One way to answer this question is to assume perturbations $\tilde{\theta}$ and $\tilde{\phi}$ about the known vortex solutions, and linearize the equations of motion in these perturbations, obtaining a pair of coupled linear differential equations for $\tilde{\theta}$ and $\tilde{\phi}$. These will be in the form of an eigenvalue problem, whose bound state solutions with negative energies will correspond to instabilities of the original unperturbed vortex. If no negative energy bound states exist then the vortex is stable. Unfortunately, the eigenvalue problem cannot be solved for the case of the planar vortex, nor even set up for the out-of-plane vortex.

An alternative is to make a numerical integrations of the equations of motion, using a continuum vortex solution discretized on a lattice as the initial condition. A Landau-Gilbert damping term can be added to the equations of motion, to take up excess energy and reduce the effects of putting a continuum solution for an infinite system on a finite sized lattice. The equations of motion become

$$\frac{d\vec{S}_i}{dt} = \vec{S}_i \times \vec{F}_i - \epsilon\, \vec{S}_i \times (\vec{S}_i \times \vec{F}_i) \qquad (16)$$

with $\vec{F}_i$ given in (3) and ϵ is the damping strength. Integrations were done[14] on a 40 x 40 lattice, with Neumann or free boundary conditions and a 4th order Runge-Kutta scheme, time step 0.04. The equations were integrated out to several hundred units with damping $\epsilon = 0.1$, starting from a planar vortex initial condition for a set of values $0 \leq \lambda \leq 1$. On a square lattice, for $\lambda < 0.72$, the planar vortex is found to be stable under these conditions (θ remains zero everywhere). For $\lambda \geq 0.72$, the simulation produces a striking destabilization of the planar vortex into an out-of-plane vortex, as seen in Figure 1. Similar results were obtained by repeating the simulations on triangular and hexagonal lattices. On the triangle lattice (coordination number $z = 6$) the planar vortex becomes unstable at $\lambda \approx 0.62$, and on the hexagonal lattice ($z = 3$) the planar vortex becomes unstable at $\lambda = 0.86$. The stability of the planar vortex is diminished by increasing coordination number and is strongly affected by the discreteness of the lattice.

A comparable set of simulations using the approximately known out-of-plane vortex as initial condition led to the same conclusion; there is a crossover value λ_c, below which only planar vortices are stable and above which only out-of-plane vortices are stable. These statements are also true for vortices in the XY <u>anti</u>ferromagnet on a square lattice, where we find $\lambda_c \approx 0.71$. These results are especially relevant for

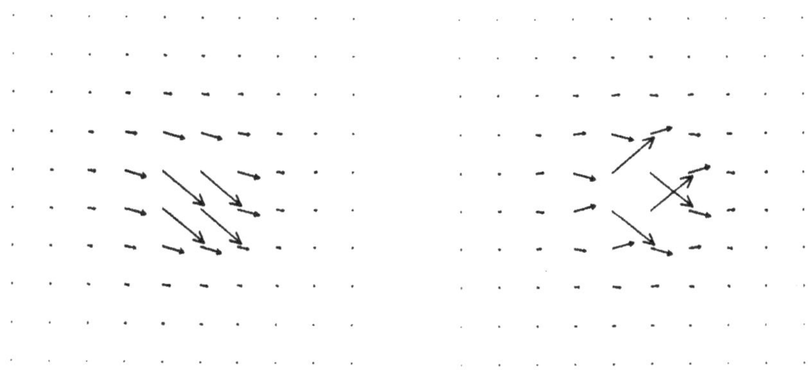

Figure 1. S^z spin components for out-of-plane vortices, with λ=0.80, for a) ferromagnet, b) antiferromagnet. The lengths of the arrows are proportional to the out-of-plane angle θ; the arrows are oriented at angle θ with respect to horizontal in the diagram. Note the antisymmetry of the antiferromagnetic vortex.

phenomenological calculations of the vortex contribution to the z-component spin-spin correlation functions.

2.5 Moving Vortices

It is useful to consider spin profiles for moving vortices, based on ideas from one-dimensional soliton dynamics. Boosting a static sG soliton for an easy-plane ferromagnet, which has $S^z = 0$, results in a profile with $S^z \sim \dot{\phi} = -v\phi_x$. The out-of-plane component is proportional to the soliton velocity v and the spatial derivative of ϕ. Generalizing to two dimensions, a planar vortex boosted along the x-direction should have

$$S^z \sim -v\phi_x = \frac{vy}{r^2} \tag{17}$$

or, for a vortex moving along an arbitrary direction, this is

$$S^z = -\frac{\vec{v} \cdot \hat{e}_\alpha}{4(1-\lambda)JSr}, \quad \text{for } r \to \infty \tag{18}$$

where (r,α) are polar coordinates. This is an approximate far-field solution. Probably effects of discreteness can strongly affect the vortex core, and an adequate $r \to 0$ solution is not known. However, it is important to note the symmetry of the moving planar vortex. It is antisymmetric about a line through the center in the direction of the velocity. Along this line S^z changes sign. The presence of nonzero z components implies that moving planar vortices will contribute to $S^{zz}(\vec{k},\omega)$.

3. NUMERICAL CALCULATIONS OF $S^{\alpha\alpha}(\vec{k},\omega)$

Kawabata and Bishop[15] introduced a convenient combined Monte Carlo-molecular dynamics technique for performing microcanonical ensemble spin dynamics calculations. First, a Monte Carlo (MC) simulation (canonical ensemble) is used to generate a set of equilibrium configurations for a chosen temperature. Then these configurations can be used in an energy-conserving numerical integration of the spin dynamics equations of motion, written most efficiently in terms of the xyz spin components.

Typical simulations[11,14,16] reviewed here involved a 100 x 100 square lattice, with periodic boundary conditions. The system was considered to have two sublattices. In the MC algorithm all spins on one sublattice were updated simultaneously, so that the algorithm could be vectorized. Individual spins were updated in one of two ways: i) by using trial spins uniformly distributed on the units sphere, or, ii) by adding small increments to the old spin position, scaled by a parameter ℓ, and adjusting the length ℓ to give an acceptance rate from 30% to 40%. The former worked well for higher temperatures, the latter was more efficient for lower temperatures but may not be recommended for extremely low temperature. Most interest was in $T/J \gtrsim 0.5$, where J is the near neighbor exchange, so low temperatures were not a problem. About 2000 passes through the lattice were used to equilibrate the system, with the temperature slowly lowered to the desired value exponentially, in the sense of simulated annealing. Then, about 10,000 passes were used to generate an initial state for the spin dynamics, and for computing equilibrium averages.

In the spin dynamics part of the simulation, a 4th order Runge-Kutta integrator with fixed time step of 0.04 was used. The spin configuration (or, selected components of its spatial Fourier transform) was sampled every m time steps, with m chosen from 1 to 32 depending on the frequencies of interest. A set of 512 samples were taken, to take advantage of a fast Fourier transform. Time-displaced correlations of the spatial Fourier transform of $\vec{S}(\vec{r},t)$, that is, $C^{\alpha\alpha}(\vec{k},\tau) = \langle S^{\alpha}(\vec{k},t)\ S^{\alpha}(\vec{k},t+\tau)\rangle$ were formed; the average is over initial times t, assuming periodic boundary conditions in time over an interval $\Delta t = 512\cdot m\cdot 0.04$. The correlation $C^{\alpha\alpha}$ was multiplied by a Gaussian window function for smoothing the effects of the finite time interval; then the product was Fourier transformed with respect to the time displacement τ, thereby producing the dynamic structure function $S^{\alpha\alpha}(\vec{k},\omega)$.

Finally, the entire process described above was repeated 3 to 5 times and the set of results were averaged together to give $S^{\alpha\alpha}(\vec{k},\omega)$. The result was then plotted and analyzed in terms of central peak and spin wave components, characterized by appropriate intensities and widths.

3.1 Numerical Results for Dynamic Correlations: XY Model

Some results for the correlations $S^{\alpha\alpha}(\vec{k},\omega)$ for ferromagnetic and antiferromagnetic XY models are shown in Figures 2 and 3. In both cases the obvious feature is development of central peak intensity with increasing temperature, which is attributed to the unbinding of vortex-antivortex pairs. At the same time a softening and broadening of the spinwave peak occurs. For the ferromagnet, the width and intensity of the central peak has been compared with predictions of the ideal gas theory.[11] For the antiferromagnet, the optical spinwave appears in the S^{xx} correlation function, and there is a broad background intensity with frequencies ranging up to the spinwave frequency, while the acoustic spinwave appears in S^{zz} (see next section).

4. ANISOTROPIC HEISENBERG ANTIFERROMAGNETS

Reconsider model (1), but with antiferromagnetic exchange:

$$H = J \sum_{(i,j)} (S_i^x S_j^x + S_i^y S_j^y + \lambda S_i^z S_j^z) \tag{19}$$

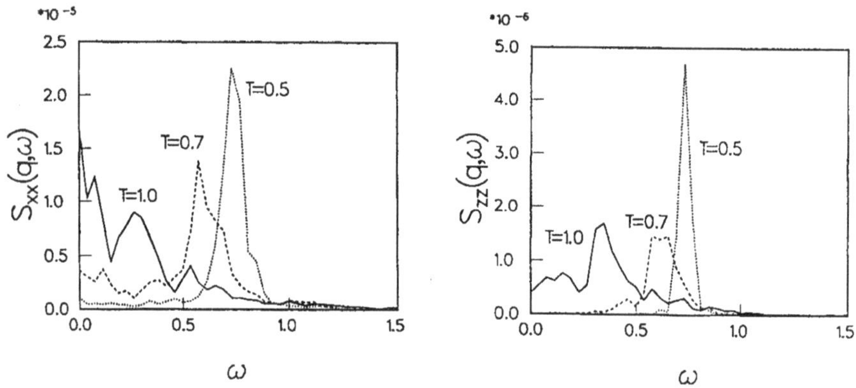

Figure 2. Correlation functions of in-plane and out-of-plane components for the XY ferromagnet at wavevector $\vec{k}a = (0.1,0.1)\pi$, from MC-MD simulations described in the text (J=1).

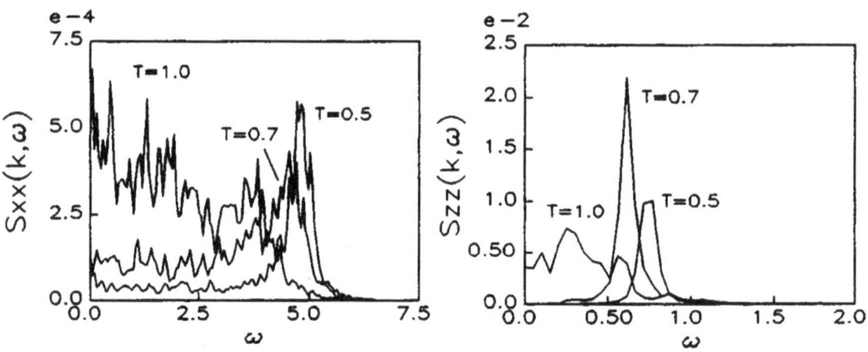

Figure 3. Correlation functions of in-plane and out-of-plane components for the XY antiferromagnet at wavevector $\vec{k}a = (0.1,0.1)\pi$, from MC-MD simulations described in the text. The acoustic spinwave appears in S^{zz}; the optical spinwave appears in S^{xx}.

with J > 0 here. The discrete equations of motion are given by equations (3) and (4) but with the spin of J reversed (equivalent to reversing the time axis). Now, the spinwaves are fluctuations from a ground state with alternating up/down spins on the two sublattices, e.g., $\vec{S}_A = S(1,0,0)$, $\vec{S}_B = S(-1,0,0)$. The spinwave dispersion then generally has two branches:

$$(\omega/2JS)^2 = (4-\lambda c^2) \pm 2(1-\lambda)c \quad ; \quad c = \cos k_x a + \cos k_y a \qquad (20)$$

with +/- corresponding to optical/acoustic branches. For $\lambda \neq 1$, the symmetry of these modes is such that only the optic mode is seen in $S^{xx}(\vec{k},\omega)$, and only the acoustic mode is seen in $S^{zz}(\vec{k},\omega)$. However, for $\lambda = 1$, the branches are degenerate and this distinction disappears.

4.1 A Continuum Limit: Heisenberg Model $\lambda=1$

A continuum limit can be made following Mikeska[6] in terms of four fields Θ, Φ, θ, ϕ, assuming fluctuations from an antiferromagnetic configuration,

$$\vec{S}_{\substack{A\\B}} = \pm\, S\, [\cos(\Theta\pm\theta)\cos(\Phi\pm\phi),\ \cos(\Theta\pm\theta)\sin(\Phi\pm\phi),\ \sin(\Theta\pm\theta)] \tag{21}$$

and then the isotropic equations of motion, assuming small spatial derivatives and $\theta \ll 1$, $\phi \ll 1$, are

$$\dot{\Theta}/JS = -8\phi\cos\Theta \tag{22}$$

$$\dot{\theta}/JS = \cos\Theta\,\nabla^2\Phi - \sin\Theta\,[2\vec{\nabla}\Theta\cdot\vec{\nabla}\Phi - 8\theta\phi] \tag{23}$$

$$\dot{\Phi}/JS = 8\theta\sec\Theta \tag{24}$$

$$\dot{\phi}/JS = \sin\Theta\,[8\theta^2\sec^2\Theta - 8\phi^2 - |\nabla\Phi|^2] - \sec\Theta\,\nabla^2\Theta\ . \tag{25}$$

For static solutions, $\theta = \phi = 0$, and the equations for Θ and Φ are identical to those for the ferromagnet (this is true for any λ). Therefore, static antiferromagnet topological excitations are the same as those in the ferromagnet. For example, we can have the "instanton" solutions[17]

$$\Phi = \tan^{-1}(y/x); \qquad \sin\Theta = \frac{w^2-r^2}{w^2+r^2} \tag{26}$$

where w is an arbitrary width parameter, and this configuration has finite energy $E = 4\pi JS^2$, independent of w and the system size! In thermal equilibrium there can be a small population of instantons in a background field dominated by spinwaves, with both making contributions[18] to $S(\vec{k},\omega)$. Theoretical models[19,20] for the finite temperature spinwave contribution have been developed, especially for wavevectors near the antiferromagnetic Bragg point $\vec{k}a = (\pi,\pi)$. For comparison, some numerical simulation results[21] are shown in Figure 4. These responses fit well to a product

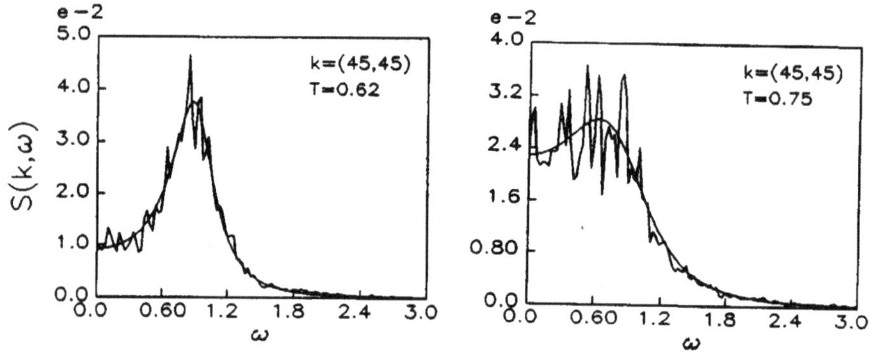

Figure 4. $S(\vec{k},\omega)$ for the Heisenberg antiferromagnet at wavevector $\vec{k}a = (45,45)\pi/50$ (from 100 x 100 lattice), from MC-MD simulations. The smooth curves are least squares fits using products of Lorentzians. Such fits account for most of the ω=0 intensity.

of Lorentzians, symmetrically located at positive and negative frequencies. Spinwave linewidths and frequencies from such fits are shown in Figure 5. Softening and broadening with increasing temperature are readily seen. These results are consistent with the dynamic scaling theory of Chakravarty et al.[19] Further details of these simulations will be published elsewhere.

5. A MODEL FOR Rb_2CrCl_4: IN-PLANE SYMMETRY-BREAKING

Hutchings et al.[3] has studied the compound Rb_2CrCl_4 using neutron scattering, and designed the following two-sublattice model Hamiltonian on a square lattice:

$$H = -J \sum_{(i,j)} \vec{S}_{i1} \cdot \vec{S}_{j2} + \sum_i D[(S^z_{i1})^2 + (S^z_{i2})^2]$$
$$- \sum_i P\,[(S^x_{i1})^2 + (S^y_{i2})^2] \qquad (27)$$

The indices 1 and 2 refer to even and odd sublattices. The model includes ferromagnetic exchange J, easy-plane anisotropy D, and most importantly, an Ising anisotropy P which is sublattice-dependent. P reflects the crystal field effects of the Cl^- ions on the magnetic Cr^{++} ions. On the "1" sublattice, the Cl^- ions are closer to the Cr^{++} ions measuring along the y-axis, and on the "2" sublattice, the Cl^- ions are closer to the Cr^{++}

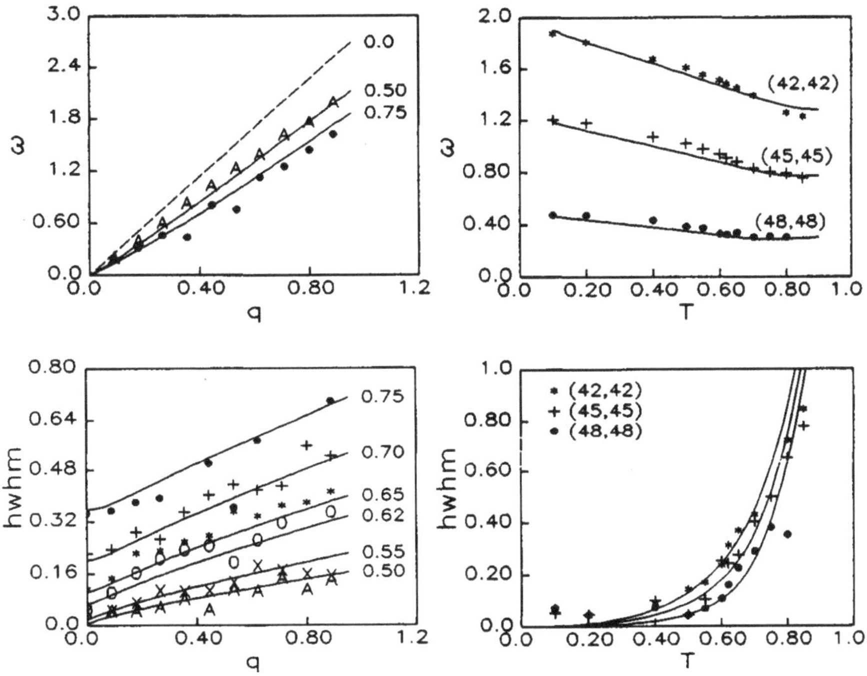

Figure 5. MC-MD data for spinwave frequencies (ω) and linewidths (hwhm) for the Heisenberg antiferromagnet, a) versus reduced wavevector $q = |\vec{k}-(\pi,\pi)|$ and b) versus temperature T. Curves in a) are labeled by various temperatures, curves in b) are labeled by wavevectors in units of $\pi/50a$. Solid curves correspond to a fit to the theory of reference 19, using parameters $\delta=2.5$, $\gamma_0=1.7$, $\mu=1.7$ and $\theta=0.7$.

ions measuring along the x-axis. Therefore the Ising axes of the two sub-lattices are perpendicular. When competing with the nearest neighbor exchange, this results in an approximate 4-fold anisotropy within the easy (XY) plane. José et al.[22] have suggested that a 4-fold in-plane symmetry breaking ($p = 4$) is a borderline case for Kosterlitz-Thouless transitions. it is expected that XY-like systems with $p < 4$ will not show a KT transition, while $p > 4$ systems will, making Rb_2CrCl_4 an interesting system to consider.

5.1 A Continuum Limit

Following the method used for antiferromagnets, spins on the two sublattices are parameterized as[16]

$$\vec{S}_{\frac{1}{2}} = S[\cos(\Theta \pm \theta)\cos(\Phi \pm \phi),\ \cos(\Theta \pm \theta)\sin(\Phi \pm \phi),\ \sin(\Theta \pm \theta)]. \tag{28}$$

Assuming small spatial derivatives and $\theta \ll 1$, $\phi \ll 1$, the resulting equations of motion have planar solutions which satisfy (with $\Theta = \theta = 0$)

$$\Phi = \pm\ \pi/4,\ \pm\ 3\pi\ 4; \qquad \phi = (P/8J)\ \sin 2\Phi. \tag{29}$$

These represent the 4 possible local ground state configurations (or, 4 possible domain orientations). Another planar solution exists, where Φ must satisfy[16]

$$\nabla^2\Phi - \frac{1}{4D'JS^2}\ \ddot{\Phi} = \frac{1}{2}\left(\frac{P}{4J}\right)^2 \sin 4\Phi \tag{30}$$

where $D' = D + P/2$. This is a time-dependent sG equation in the variable 4Φ. A static vortex-like solution given by Hudak,[23] is a synthesis of a vortex circulation constrained by the 4-fold domain pattern at large distances (see Figure 6),

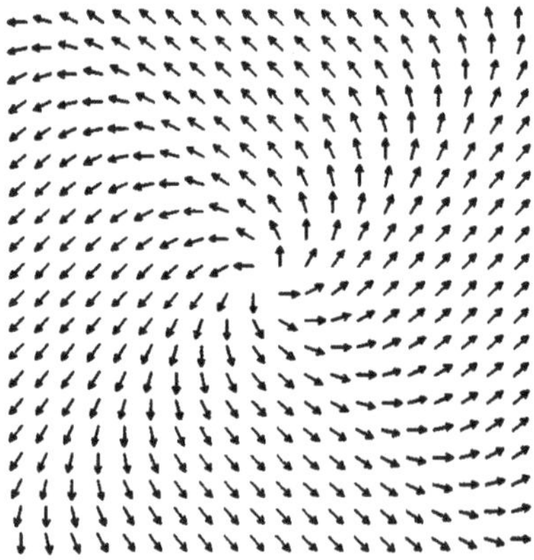

Figure 6. The 4-fold-symmetric domain-vortex of Equation (31), for $P/J = 1.0$, $n = 1$.

$$\Phi = \pm \tan^{-1} \{ \sinh[\frac{P}{2J} (y-y_o)]/\sinh [\frac{P}{2J} (x-x_o)]\} + n\pi/4 \qquad (31)$$

where n is an odd integer. This configuration, consisting of four domains separated by $\pi/2$ walls, has an energy which is linearly proportional to the system size (i.e., proportional to the length of the walls), in contrast to the logarithmic size dependence for the vortex in the XY model. However, a vortex-antivortex pair has a finite energy which is linearly proportional to the separation. As a result entropy and energy arguments for transitions in this system could be considerably modified from the XY model results. There are some numerical simulation data available from Gouvea et al.,[16] but the effects of the degree of symmetry-breaking and the freedom of out-of-plane spin components on KT transitions has not been clarified.

6. SUMMARY AND OUTLOOK

The rich variety of possible magnetic models in two dimensions gives remarkable opportunities for investigating dynamics of nonlinear excitations, including domain walls, vortices, instantons, and their subsequent effects on spinwaves. In addition, in-plane anisotropy and applied fields modify the vortex shape and dynamics, for example, as in Rb_2CrCl_4. Anisotropies and out-of-plane spin components are expected to modify vortex unbinding transitions but the details are not understood.

In the anisotropic Heisenberg models (both signs of J), there are planar and out-of-plane vortices; for a given anisotropy strength λ only one type will be stable. The planar vortex is unstable for λ greater than a crossover value λ_c, and the out-of-plane vortex is unstable for $\lambda < \lambda_c$.

In the Hutchings model for Rb_2CrCl_4, the competition between exchange and the sublattice-dependent Ising anisotropy leads to a special vortex-like excitation which might be called a "domain-vortex." It is a 4-domain coherent structure terminated in a singular point with a vortex core.

Further understanding of these and related systems will act as a foundation for investigating coherent dynamical modes, such as those involved in driven systems and in pattern formation.

Acknowledgement

The hospitality of Universidade Federal de Minas Gerais, where I had many beneficial discussions with M.E. Gouvêa and A.S.T. Pires, is greatly appreciated.

REFERENCES

1. Steiner, M., Kakuri, K. and Kjems, J.K., Z. Phys. B53, 117 (1983); Boucher, J.P., Regnault, L.P., Rossat-Mignod, J., Renard, J.P., Bouillot, J., Stirling, W.G. and Mezei, F., Physica 120B, 241 (1983).
2. Regnault, L.P. and Rossat-Mignod, J., in Magnetic Properties of Layered Transition Metal Compounds, editors de Jongh, L.J. and Willett, R.D. (Reidel, Dordricht, 1987).
3. Hutchings, M.T., Als-Nielsen, J., Lindgard, P.A. and Walker, P.J., J. Phys. C: Solid State Phys. 14, 5327 (1981).
4. Elahy, M. and Dresselhaus, G., Phys. Rev. B30, 7225 (1984).
5. Mikeska, H.J., J. Phys. C11, L29 (1978).
6. Mikeska, H.J., J. Phys. C13, 2913 (1980).
7. Bishop, A.R. and Schneider, T. (eds.) Solitons in Condensed Matter Physics, Springer-Berlin (1978).
8. Currie, J.F., Krumhansl, J.A., Bishop, A.R. and Trullinger, S.E., Phys. Rev. B22, 477 (1980).
9. Kumar, P., Phys. Rev. B25, 483 (1982); Physica 5D, 359 (1982); Magyari, E. and Thomas, H., Phys. Rev. B25, 531 (1982); J. Phys. C16, L535 (1983).
10. See Nonlinearity in Condensed Matter, edited by Bishop, A.R., Campbell, D.K., Kumar, P. and Trullinger, S.E., p. 24-80, Springer-Verlag (1987).
11. Mertens, F.G., Bishop, A.R., Wysin, G.M. and Kawabata, C., Phys. Rev. Lett. 59, 117 (1987), Phys. Rev. B39, 591 (1989).
12. Kosterlitz, J.M. and Thouless, D.J., J. Phys. C: Solid State Phys. 6, 1181 (1973); in Progress in Low Temperature Physics, edited by Brewer, D.F., Vol. 7B, Chap. 5 (North-Holland, Amsterdam, 1978).

13. Hikami, S. and Tsuneto, T., Prog. Theo, Phys. 63, 387 (1980).
14. Gouvêa, M.E., Wysin, G.M., Bishop, A.R. and Mertens, F.G., Phys. Rev. B39, 11840 (1989); in Computer Simulation Studies in Condensed Matter Physics, edited by Landau, D.P., Mon, K.K. and Schüttler, H.B., Springer-Verlag (1988).
15. Kawabata, C. and Bishop, A.R., Solid State Commun. 42, 595 (1982); ibid, 60, 169 (1986).
16. Gouvêa, M.E., Wysin, G.M., Bishop, A.R. and Mertens, F.G., J. Phys.: Condens. Matter 1, 4387 (1989).
17. Trimper, S., Phys. Lett. 70A, 114 (1979), Belavin, A.A. and Polyakov, A.M., Pis'ma Zh. Eksp. Torr. Fiz. 22, 503 (1975) [JETP Lett. 22, 245 (1975)].
18. Kawabata, C. and Bishop, A.R., Solid State Commun. 33, 453 (1980).
19. Tyc, S., Halperin, B. and Chakravarty, S., Phys. Rev. Lett. 62, 853 (1989); Chakravarty, S., Halperin, B. and Nelson, D (unpublished).
20. Becker, T. and Reiter, G. (unpublished).
21. Wysin, G.M. and Bishop, A.R. (unpublished).
22. José, J.V., Kadanoff, L.P., Kirkpatrick, S., Nelson, D.R., Phys. Rev. B16, 1217 (1977).
23. Hudak, O., Phys. Lett. 89A, 245 (1982).

Spreading of Damage in Ising Models

Uriel M. S. Costa†, Heber R. da Cruz† and Evaldo M. F. Curado*

†Departamento de Física, Universidade Federal de Alagoas,
Cidade Universitária, 57.061 Maceió – Al, Brazil.
*Centro Brasileiro de Pesquisas Físicas
Rua Xavier Sigaud, 150, Urca. 22.290 Rio de Janeiro – RJ, Brazil

ABSTRACT

Here we review the work done on the spreading of damage in a cooperative system. The question "How does a perturbation spread throughout a cooperative system, composed of interacting subunits?" is answered for the Ising ferromagnet and for the Ising spin glass, using Glauber and heat bath dynamics, in a square net and simple cubic lattice. It is found that spreading of damage is greatly hindered below a threshold temperature T_s when the Glauber dynamics is used. When heat bath dynamics is used, however, three temperature regimes are found. In most cases this temperature is very close to the usual critical temperature.

1. INTRODUCTION

The question of how a perturbation spreads on a cooperative system, with interacting subunits, has risen interest from researchers working in a variety of fields. In genetics, for example, one can ask what a single mutation in a gene can cause to the whole animal. This study has been done on cellular differentiation[1)]. In computer science, the consequences of a software or hardware error, in a computer, may or may not bring the whole system to a halt.

Unable to answer the above questions, physicists turned back to a much simpler problem, namely to understand the consequence of a small "damage" (spin flip) in Ising and cellular automata systems[2–5)]. Here we intend to summarize the most recent papers studying the spreading of damage in Ising models including Ising spin glass, drawing some conclusions on the results so far obtained.

This paper is organized as follows. In section 2 we will define the different dynamics used. In section 3 we will describe how we follow the spreading of damage in a system. In section 4 we will discuss the spreading of damage in the Ising ferromagnet in two and three dimensions. In section 5 we will present the same study in an Ising spin glass. Finally, in section 6 we will summarize our conclusions.

2. ISING MODEL DYNAMICS

In the heat bath dynamics a spin becomes up with probability

$$p_i^{HB}(t) = [1 + e^{-2\beta h_i(t)}]^{-1} \tag{1}$$

where $h_i(t) = \sum_j J_{ij} S_j + H$ is the local field, $\beta=1/k_B T$ and H is the magnetic field.

In the Glauber dynamics, a spin is flipped with probability

$$p_i^{G}(t) = [1 + e^{2\beta S_i h_i(t)}]^{-1} \tag{2}$$

where $h_i(t)$ is defined above. The factor $2S_i h_i(t)$ in the exponent of the above expression represents the energy difference between the old and new configurations. Naturally, when the energy difference is negative (the spin is in the "wrong" direction) the probability of flipping is high, and when the energy difference is positive the probability of flipping is low.

In terms of probability there is no difference between this and the Glauber dynamics. This can be easily seen from eq. (2). When S_i is -1 eq. (2) will be identical to eq. (1). When S_i is $+1$ eq. (2) will give the probability that this spin becomes -1, therefore we should compare one minus this probability with eq. (1) and we will see again that they are identical.

The difference lies in the use of the random numbers as we shall explain with an example. Let us assume that for a given site $h_i(t)$ is the same in both configurations but in configuration A S_i is up and in configuration B S_i is down. Let z be a random number. In the heat bath dynamics if $z>p_i$ the spin S_i in both configurations will be up at $t+1$, and both down if $z\leq p_i$. Therefore they will inevitably become equal. In the

Glauber dynamics the spin S_i in configuration A will be flipped (and therefore become down) if $z>p_i$ and the spin S_i in configuration B will be flipped (and therefore become up) if $z>1-p_i$. There are then three possibilities: a) $z>max(p_i,1-p_i)$, both flip and remain different; b) $max(p_i,1-p_i)>z>min(p_i,1-p_i)$, one flips and the other does not and they become equal; c) $z<min(p_i,1-p_i)$, none flip and remain different.

In the Metropolis dynamics a spin is flipped with probability

$$p_i(t) = min\left(1, e^{-2\beta S_i h_i(t)}\right)$$

Although the Metropolis dynamics uses probability different from the Glauber dynamics, in the end they are only slightly different. We can see that when the energy difference is negative the probability of flipping is one and when the energy difference is positive the probability of flipping is less than one. In all cases known to us the difference in the results obtained by the Metropolis and the Glauber dynamics and essentially the same.

The Q2R dynamics is a cellular automaton. The deterministic rule is that a spin is flipped if and only if the energy cost is zero (i.e., half of the neighbor of a spin are up and half are down).

3. THE SPREADING OF DAMAGE ALGORITHM

The study of the spreading of damage in Ising models consists in studying the temporal evolution of two configurations which differ by 1, 2 or even half of the sites, at time t=0. As the Ising model has no intrinsic dynamics an "external" is needed. To study the temporal evolution of such models Glauber, heat bath or Q2R dynamics have been used.

The spreading of damage simulation is done in the following manner:

(i) Take a configuration A in thermal equilibrium at a given temperature T.

(ii) Make a copy B of that configuration and introduce a damage by changing the state of the spin in one or more sites.

(iii) Treat both configurations in exactly the same manner by Apply the same algorithm and the same random numbers.

(iv) Study the time evolution of the damage.

(v) Take the configurational average by repeating the process for several different initial configurations.

(vi) Repeat the process for several values of the temperature.

The quantities that we are interested in this work are:

(a) Hamming distance: the fraction of the sites which are different in the two lattices at a given time.

(b) Total damage time: the time required for all sites to be damaged at least once.

4. ISING FERROMAGNET

In 1987 Stanley *et al.*[2] studied the spreading of damage in a 2d Ising ferromagnet in a square net, using two dynamics: Glauber and Q2R. In the latter a spin is flipped if and only if this will not change the energy.

They found, for both dynamics, that the spreading of damage is hindered below a threshold temperature T_s which agrees numerically with the Curie temperature. With the Glauber dynamics T_s is found as a sharp transition point while with Q2R they found a kinetic decay.

They measured, among others, the following quantities:

(i) τ – time required for the damage to reach the upper or lower limit of the net.

(ii) M_τ – total number of sites damaged at time $t = \tau$.

(iii) Order parameter Ψ – the fraction of damaged sites, with the initial damage infinitesimally small extrapolated for the $L \to \infty$ limit.

(iv) T_1 – time required for all spins to be damaged at least once (total damage time).

In that paper they consider three configurations. Configuration A, which is the original, undamaged one. Configuration B, which differs from A by 8 regularly distributed sites. Configuration C which differs from A by 16 regularly distributed sites, including those 8 sites from configuration B. From the time evolution of those three configurations they obtain $M_\tau(A,B)$, $M_\tau(A,C)$ and $M_\tau(B,C)$, for $t \simeq 10^3$. The order parameter Ψ is defined as:

$$\Psi = M_\tau(A,B) + M_\tau(B,C) - M_\tau(A,C)$$

Figure 1a shows the results obtained by Stanley *et al.*[2] for the order parameter ψ as a function of temperature for the Glauber dynamics while figure 1b shows the same order parameter as a function of energy for the

Q2R dynamics. Figure 1c shows the inverse total damage time versus the inverse temperature for the Glauber dynamics. All figures show that the temperature T_s cannot be distinguished from the 2d Ising ferromagnet critical temperature T_c, exhibiting a close relationship between the dynamic and static transitions.

One of us[3] studied the spreading of damage in a 3d Ising model, using Metropolis dynamics. Two temperature regimes were found, but the threshold temperature does not exactly match the critical temperature of the 3d Ising ferromagnet. The threshold temperature found ($T_s \simeq 0.96\ T_c$) numerically coincides with the critical temperature for the site percolation with correlation for the Ising model[4]. Although one transition is a dynamic effect and the other a static one, he suggested that studying this model

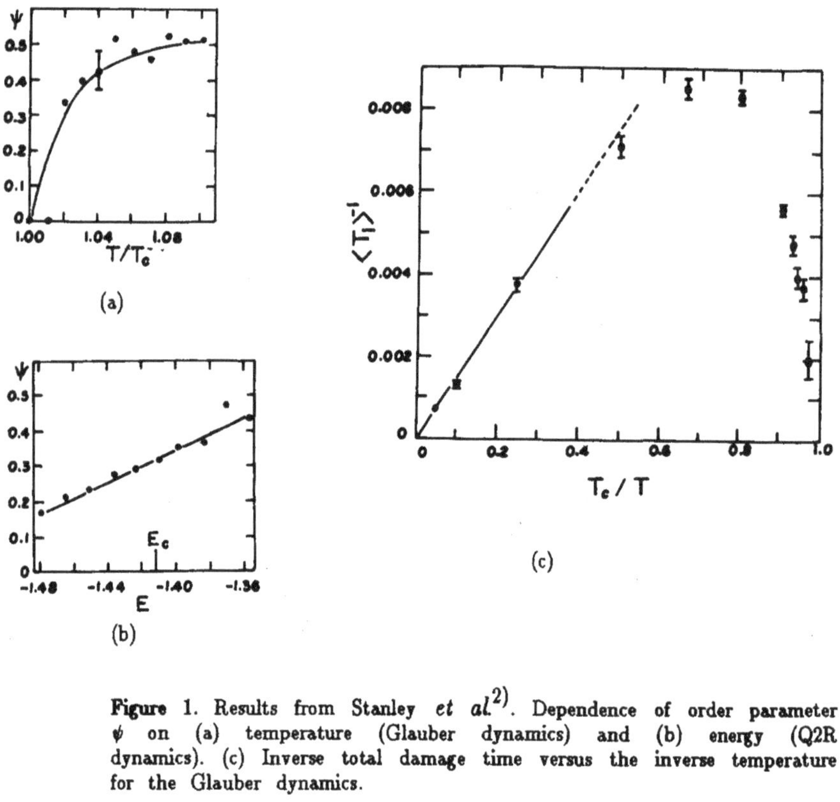

Figure 1. Results from Stanley *et al.*[2]. Dependence of order parameter ψ on (a) temperature (Glauber dynamics) and (b) energy (Q2R dynamics). (c) Inverse total damage time versus the inverse temperature for the Glauber dynamics.

with an applied magnetic field would be useful to check whether there is any connection between the spreading of damage and correlated percolation.

On the same line Le Caër[5] studied the temporal evolution of two cubic ferromagnetic systems differing by one spin only at $t=0$, for different temperatures and different values of a uniform magnetic field in the z direction. He used the Glauber dynamics which, as noted, is slightly different from the Metropolis one. At zero field he finds a chaotic regime for $T > T_s \simeq 0.91\ T_c$ (Fig. 2a). It seems that the use of different dynamics lead to, at least, quantitatively different results. This is possibly at least one of the reasons for the discrepancy of the temperatures found by these two authors. The specific explanation for this case may be that when Metropolis dynamics is used equilibrium is reached in less time, therefore facilitating the spreading of damage.

Defining a parameter that measures the correlation between the two configurations he found a "correlation" transition at T_c. For $T>T_c$ the two configurations are not correlated while below T_c they are. In the presence of a magnetic field the damage spreads for $B<B_c$. The magnetization obtained fits well to $M(B_c,T) = 1 - \exp(T_c-T)$ for $4.4 \lesssim T \lesssim 40$, as one can see by comparing the results obtained from simulation with the one obtained from this expression for B_c (Fig. 2b) and the concentration of minority spins (not shown). He also verified that there is no relation between the site percolation with correlation problem and spreading of damage in the 3d Ising ferromagnet.

Some exact relations between static (thermodynamic) and dynamic transitions have recently been derived by Coniglio *et al.*[6] and Mariz[7] using the heat bath dynamics. Coniglio *et al.* showed the connection between the damage in an Ising ferromagnet and thermal properties (magnetization, susceptibility, for example). Mariz generalized these results for the Potts and Ashkin–Teller models, deriving exact relations between specific combinations of damages and macroscopic thermal quantities. Exact relations between the damage an thermal properties in the Glauber dynamics are still lacking.

5. ISING SPIN GLASS

In a recent paper Derrida and Weisbuch[8] studied the temporal evolution of the Hamming distance of two configurations of the 3d Ising

(a)

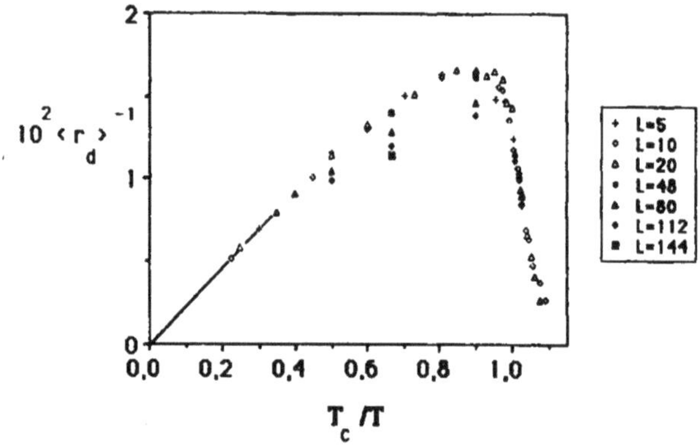

(b)

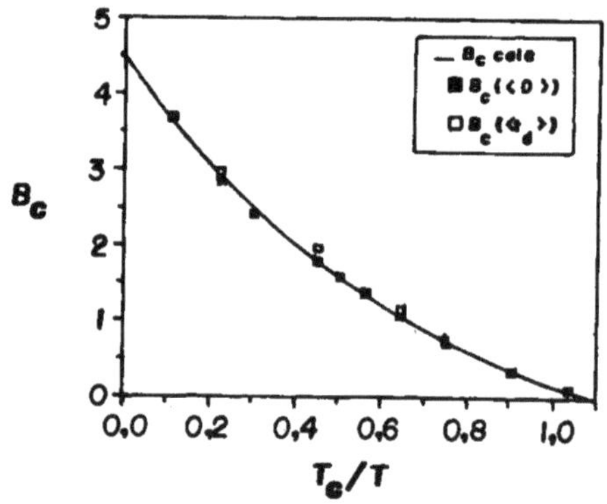

Figure 2. Results from Le Caër[5]. (a) Inverse scaled time $\langle r_d \rangle^{-1}$ as a function of T_c/T, with $r_d = t_d(L/10)^{0.67}$. (b) Critical field B_c as a function of temperature T for L=10, $B_c(\langle D \rangle)$, sequential updating (full squares); L from 48 to 112, $B_c(\langle t_d \rangle)$, simultaneous updating (open squares); B_c calculated from $M(B_c,T) = 1 - \exp(T_c - T)$ (full curve).

spin glass submitted to the same thermal noise. They used the heat bath dynamics.

They study three initial conditions: (a) initial Hamming distance 1 (all sites different); (b) initial Hamming distance 1/2 (random configurations); (c) initial Hamming distance 1/N (one site different). When measuring the Hamming distance after time ($t \lesssim 500$, for L=8, 12) they find three temperature regimes (see Fig. 3):

- a high temperature regime with $T>T_1$ ($T_1 \simeq 4.1J$) where the Hamming distance becomes zero independent of the initial conditions.
- an intermediate regime $T_2<T<T_1$ ($T_2 \simeq 1.8J$) where the Hamming distance is non–zero and independent of the initial conditions.
- a low temperature regime $T<T_2$ where the Hamming distance is dependent of the initial conditions.

For the same model, some very recent results from de Arcangelis *et al.*[9] studied the distance between states and the Edwards–Anderson parameter and verified that their results are consistent with a multi–valley structure for the phase space of the Ising spin glass below $T_{sg} \simeq 1.9J$ (the

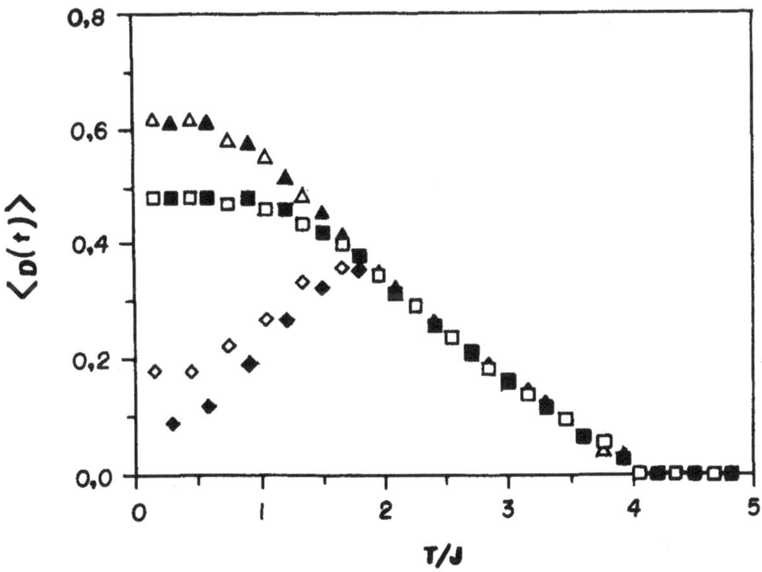

Figure 3. Results from Derrida and Weisbuch[8]. Hamming distance versus temperature. White signs correspond to L=8 and black signs to L=12. Triangles are for initially opposed configurations (case a), squares for random configurations (case b), and diamonds for configurations differing initially by only one spin (case c).

distance between the states depends on the initial damage). The same multi–valley structure is also obtained by applying the magnetic field.

da Cruz and collaborators[10)] applied Glauber dynamics to the 3d Ising spin glass and found, contrary to what was found using the heat bath dynamics, only two temperature regimes (Fig. 4a, 4b):

i) $T > T_s < T_{sg} = 1.2J$ where the damage propagates freely, corresponding to the chaotic phase.

ii) $T < T_s$ where the damage spreading is hindered, corresponding to the frozen phase.

A simple cubic lattice (L=10) was used and the time evolution was followed for times up to $t \simeq 5{,}000$ time steps. In this model it was not clear the connection between the thermodynamic transition and the spreading of damage.

In order to try to understand the relation between the spreading of damage and the thermodynamic phase transition the same authors decided to study the spreading of damage in a system which was known not to have a thermodynamic phase transition, namely the 2d Ising spin glass, in a square

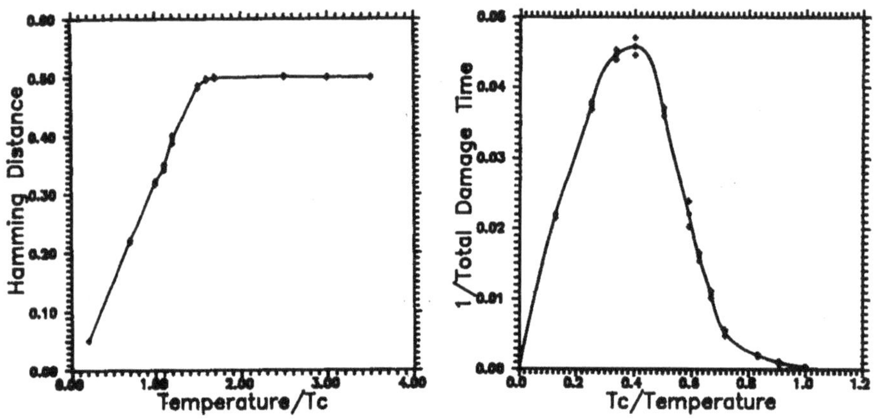

(a) (b)

Figure 4. Results from da Cruz *et al.*[10)]. Both results are for the 3d Ising spin glass. (a) Hamming distance plotted against temperature. (b) Inverse total damage time plotted against inverse temperature.

net. Contrary to what would be expected two temperature regimes were found, just like in the 3d case (Fig. 5a, 5b). We believe that a computational effort must be made in order to verify if this 2d phase persists after a very long time.

5. CONCLUSIONS

From the papers described above we can draw the following conclusions:

(i) the spreading of damage is dependent of the dynamic used.

(ii) for all known cases, when there is a thermodynamic phase transition there is a spreading of damage transition (chaotic–frozen) at a nearby temperature.

(iii) a spreading of damage transition can also occurs without a thermodynamic phase transition (at least for the observation times used).

(iv) exact relations between spreading of damage (using heat bath dynamics) and thermodynamic properties exist.

(v) Results are, at least qualitatively, insensitive to: making the damage before thermal equilibrium is reached and making the update sequentially or simultaneously (multispin coding)[11].

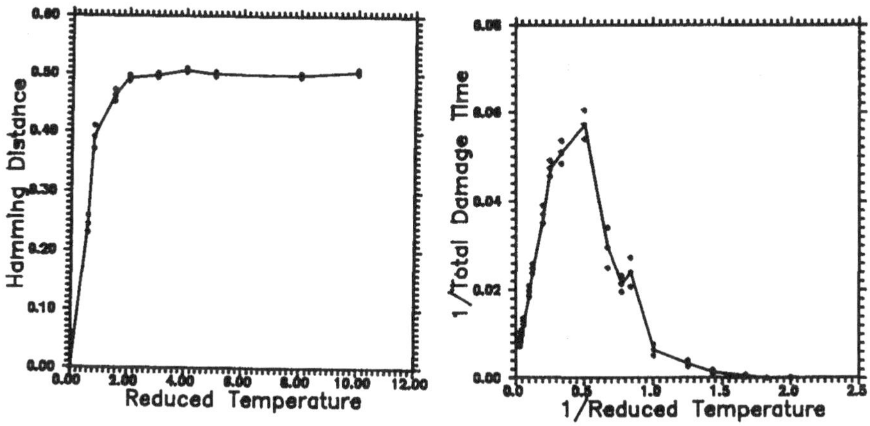

(a) (b)

Figure 5. Results from da Cruz *et al.* Both results are for the 2d Ising spin glass. (a) Hamming distance plotted against temperature. (b) Inverse total damage time plotted against inverse temperature.

(vi) In all known cases the results do not change qualitatively with the cell size [2,3,4,5,8,9].

(vii) There are systems (spin glasses, for example) where the final damage depends on the initial one. This behavior is observed with both Glauber[10] and heat bath[8,9] dynamics.

REFERENCES

1. Kauffmann, S. A., *J. Theor. Biol.* 22, 437 (1969). Kauffmann S. A., *Physica D* 10, 145 (1984).
2. Stanley, H. E., Stauffer, D., Kertész, J. and Herrmann, H. J. *Phys. Rev. Lett.* 59, 2326 (1987).
3. Costa, U. M.S., *J. Phys A* 20, L583 (1987).
4. Heermann, D. W. and Stauffer, D., *Z. Phys. B* 44, 339 (1981).
5. Le Caër, G., *J. Phys. A* 22, L647 (1989). Le Caër, G., preprint (1989).
6. Coniglio, A., de Arcangelis, L., Herrmann, H. J. and Jan, N., *Europhys. Lett.* 8, 315 (1989).
7. Mariz, A. M. (1989), preprint.
8. Derrida, B. and Weisbuch, G., *Europhys. Lett.* 4, 657 (1987).
9. de Arcangelis, L,, Coniglio, A. and Herrmann, H. J. (1989), preprint. de Arcangelis, L., Herrmann, H. J. and Coniglio, A. (1989), preprint.
10. da Cruz, H. R., Costa, U. M.S. and Curado, E. M. F., *J. Phys. A* 22, L651 (1989).
11. Stauffer, D. private communication.

LATTICE-GAS APPROACH TO NONEQUILIBRIUM HYDRODYNAMICS

Michel Droz *)
Department of Theoretical Physics
University of Geneva
CH-1211 Geneva 4, Switzerland

ABSTRACT:

The problem of using lattice-gas models to simulate thermal effects in a fluid is briefly reviewed. A particular model describing heat propagation in a fluid at rest is discussed. Numerical results, obtained on a special purpose computer, are compared with analytical predictions, both for the thermal conductivity and the (equilibrium and nonequilibrium) correlation functions.

1. INTRODUCTION.

In the recent years, a large effort has been devoted to the study of lattice-gas models simulating real fluids flows [1] , [2] , [3] . These approaches have been quite successful in describing two and three dimensional flows around obstacles for low Reynolds numbers.

However, one is often interested in more complex situations. One example is the case in which thermal effects are presents. Several new phenomenas can then take place as for example Rayleigh-Benard instabilities. Even for a fluid at rest, there are many interesting problems to study in the presence of a temperature gradient. One example is the study of the fluctuations in such a nonequilibrium situation.

This is a difficult problem in the usual framework of the fluctuating hydrodynamics. It is thus legitimate to investigate if generalized lattice-gas models can correctly describe such thermal effects, and provide an easier alternative approach.

*) Supported by the Swiss National Science Foundation.

The paper is organized as follows. In Section 2, we describe a two speed lattice-gas model suitable to describe the heat transport in a fluid at rest. We show briefly how to derive the thermo-hydrodynamic equations from the microdynamics ones. We then compare the consequences of these equations with the exact numerical results obtained by solving the model on a special purpose machine. In Section 3, we discuss the problem of the definition of a good temperature scale in such a model. Finally, in Section 4, we consider the nonequilibrium two-point correlation functions for the hydrodynamic variables. We compare the numerical results obtained for the lattice-gas model with the ones obtained in the framework of the fluctuating hydrodynamic. A qualitative agreement is found between the two approaches.

2. A TWO-SPEED LATTICE-GAS MODEL

Hydrodynamics is a field which can be described at two levels of reality [4] . On a microscopic level, a fluid consists of discrete particles which dynamics is given by deterministic time-reversal invariant equations of motion. On the other hand, on macroscopic level, the fluid is described by a set of irreversible thermo-hydrodynamical equations. The goal of the nonequilibrium statistical mechanics is to connect these two levels of description.

Lattice-gas models (often called cellular automata (CA) approach) provides a way to achieve this program for discrete models. The model we shall discuss has been proposed by Chopard and Droz [5] and is a two-speed HPP model. It consists in an assembly of particles travelling on a square lattice. All the particles are assumed to have a mass equal to one but may have two possible values of their kinetic energy.

The slow particles move along the horizontal and vertical directions with velocity v (cold particles), while the fast particles (hot particles) move along the diagonals with velocity $w = \sqrt{2}v$ (see Figure 1). Only one particle per direction is allowed at a given site and at a given time. The interactions between the particles are depicted in Figure 2. During the collisions, the particle number, momentum and energy are conserved. With τ the time step and λ, the lattice spacing, one has

$$v = \frac{\lambda}{\tau} \qquad w = \frac{\sqrt{2}\lambda}{\tau}$$

In view of deriving the hydrodynamical equations associated with this model, we first write the exact microdynamical equations of motion. We introduce the particle occupation numbers $n_i(\vec{r},t)$ and $m_i(\vec{r},t)$. They are the numbers

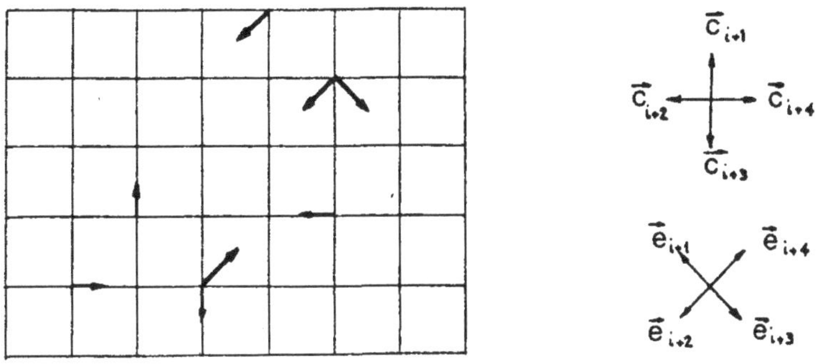

Fig. 1: Typical configuration of the lattice-gas at a given time. The slow particles move along the lattice links (directions $\vec{c}_i$, $i = 1,2,3,4$) while the fast ones move along the diagonals. (directions $\vec{e}_i$, $i = 1,2,3,4$)

of particles entering the site $\vec{r}$ at time t with velocity respectively $\vec{v}_i = v\vec{c}_i$ and $\vec{w}_i = w\vec{e}_i$ (see Figure 1).

As the model accomodates for only one particle with a given velocity at site $\vec{r}$ and time t (exclusion principle), n_i and m_i can only take the values 0 or 1. According to the collision rules defined in Figure 2, the equations of motion for n_i and m_i are, for $i = 1$ to 4, mod 4:

$$\begin{aligned} n_i(\vec{r} + \lambda\vec{c}_i; t + \tau) &= n_i(\vec{r}; t) + A_i(\{n_j(\vec{r}, t)\}, \{m_j(\vec{r}, t)\}) \\ m_i(\vec{r} + \sqrt{2}\lambda\vec{e}_i; t + \tau) &= m_i(\vec{r}, t) + B_i(\{n_j(\vec{r}, t)\}, \{m_j(\vec{r}, t)\}) \end{aligned} \tag{2.1}$$

where

$$\begin{aligned} A_i =& n_{i+1}n_{i+3}\bar{n}_i\bar{n}_{i+2} - n_i n_{i+2}\bar{n}_{i+1}\bar{n}_{i+3} \\ &+ n_{i+2}m_{i+3}\bar{n}_i\bar{n}_{i+1}\bar{n}_{i+3}\bar{m}_i\bar{m}_{i+1}\bar{m}_{i+2} - n_i m_{i+2}\bar{n}_{i+1}\bar{n}_{i+2}\bar{n}_{i+3}\bar{m}_i\bar{m}_{i+1}\bar{m}_{i+3} \\ &+ n_{i+2}m_i\bar{n}_i\bar{n}_{i+1}\bar{n}_{i+3}\bar{m}_{i+1}\bar{m}_{i+2}\bar{m}_{i+3} - n_i m_{i+1}\bar{n}_{i+1}\bar{n}_{i+2}\bar{n}_{i+3}\bar{m}_i\bar{m}_{i+2}\bar{m}_{i+3} \end{aligned} \tag{2.2}$$

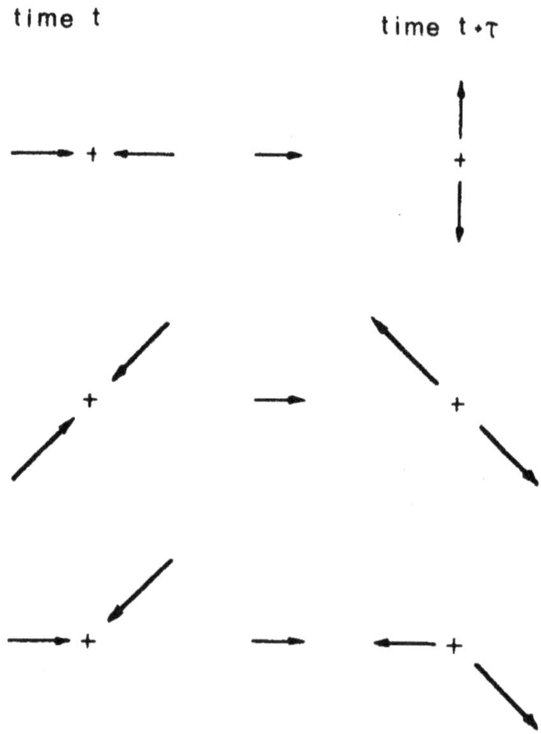

Fig. 2: Two-particles collisions allowed in the two-speed model. They are respectively between slow-slow, fast-fast and slow-fast particles. For all other situations (not trivially related by symmetry), no collision occurs.

$$
\begin{aligned}
B_i = & m_{i+1} m_{i+3} \bar{m}_i \bar{m}_{i+2} - m_i m_{i+2} \bar{m}_{i+1} \bar{m}_{i+3} \\
& + m_{i+1} n_i \bar{n}_{i+1} \bar{n}_{i+2} \bar{n}_{i+3} \bar{m}_i \bar{m}_{i+2} \bar{m}_{i+3} - m_i n_{i+2} \bar{n}_i \bar{n}_{i+1} \bar{n}_{i+3} \bar{m}_{i+1} \bar{m}_{i+2} \bar{m}_{i+3} \\
& + m_{i+3} n_{i+1} \bar{n}_i \bar{n}_{i+2} \bar{n}_{i+3} \bar{m}_i \bar{m}_{i+2} \bar{m}_{i+1} - m_i n_{i+3} \bar{n}_i \bar{n}_{i+1} \bar{n}_{i+2} \bar{m}_{i+1} \bar{m}_{i+2} \bar{m}_{i+3}
\end{aligned}
\tag{2.3}
$$

with $\bar{n}_i = 1 - n_i$ and $\bar{m}_i = 1 - m_i$

The associated lattice Boltzmann equation is obtained by a microcanonical average [5]. In terms of the average occupation numbers

$$N_i(\vec{r},t) = \langle n_i(\vec{r},t)\rangle \ , \ M_i(\vec{r},t) = \langle m_i(\vec{r},t)\rangle \tag{2.4}$$

and assuming weak correlations between the N_i's and M_i's , one obtains the following discrete Boltzmann equations

$$\begin{aligned} N_i(\vec{r}+\lambda\vec{c}_i;t+\tau) &= N_i(\vec{r},t) + A_i(N,M) \\ M_i(\vec{r}+\sqrt{2}\lambda\vec{e}_i;t+\tau) &= M_i(\vec{r};t) + B_i(N,M) \end{aligned} \tag{2.5}$$

A continuous Boltzmann equation is obtained using the so-called multi-scale expansion [5]. The density ρ, the velocity field $\vec{u}$ and the local energy per particle ϵ can be defined in terms of the N_i and M_i.

$$\rho(\vec{r},t) = \sum_{i=1}^{4} N_i(\vec{r},t) + M_i(\vec{r},t) \tag{2.6}$$

$$\rho(\vec{r},t)\vec{u}(\vec{r},t) = \sum_{i=1}^{4} \vec{v}_i N_i(\vec{r},t) + \vec{w}_i M_i(\vec{r},t) \tag{2.7}$$

and

$$\rho(\vec{r},t)\epsilon(\vec{r},t) = \sum_{i=1}^{4} \frac{\vec{v}_i^{\,2}}{2} N_i(\vec{r},t) + \frac{\vec{w}_i^{\,2}}{2} M_i(\vec{r},t) \tag{2.8}$$

The temperature is defined through the equipartition theorem and reads:

$$kT(\vec{r},t) = \frac{v^2}{2}T(\vec{r},t) = \epsilon(\vec{r},t) \tag{2.9}$$

where k plays the role of the Boltzmann constant. Note that from this definition, the temperature ranges from 1 to 2. We shall come back to this point later on. The Boltzmann equation is solved using the Chapman-Enskog procedure, and the solution allows to obtain the thermo-hydrodynamical equations. In a stationary state, and for a fluid at rest (which is the situation we shall consider), the thermo-hydrodynamic equations reduce to [5]:

$$\nabla p(\vec{r}) = 0 \quad \text{i.e.} \quad p = cte \tag{2.10}$$

and

$$\mathrm{div}(\vec{J}_T(\vec{r})) = 0 \tag{2.11}$$

where the pressure p is the diagonal part of the momentum tensor

$$\Pi_{\alpha\beta} = \sum_{i=1,4} [N_i v_{i\alpha} v_{i\beta} + M_i w_{i\alpha} w_{i\beta}] \tag{2.12}$$

and the heat current $\vec{J}_T$ is defined as follows

$$\vec{J}_T = \frac{v^2}{2} \sum_{i=1,4} [N_i \vec{v}_i + 2M_i \vec{w}_i] \tag{2.13}$$

These quantities are related to the density and temperature as follows [5]:

$$p(\vec{r}) = \frac{v^2}{2} \rho(\vec{r}) T(\vec{r}) \tag{2.14}$$

which is the well known equation of state for a classical ideal gas. And

$$\vec{J}_T(\vec{r}) = -\kappa(p,T) \nabla T(\vec{r}) \tag{2.15}$$

where the thermal conductivity $\kappa(p,T)$ has an explicit temperature dependant form, given in terms of the microscopic parameters of the model [5]

In this paper, we consider a nonequilibrium situation in which an external temperature gradient is imposed to the gas enclosed in a rectangular box. At the microscopic level, this gradient is produced by appropriate boundary conditions on the cellular automata model. More explicitly, two parallel plates (aligned with the horizontal walls of the box) are used, a cold one and a hot one. For the sites of the automaton belonging to the plates, the evolution rules are modified as follows: on the hot plate, a pair of incoming slow particles with opposite momentum is transformed into a zero momentum pair of fast particles. On the cold plate, the reverse process takes place. The magnitude of the temperarure gradient can be adjusted by varying the number of cold and hot sites on the plates.

Numerical simulations have been performed on this model [5] and the results are in good agreement with the theoretical predictions following from equations (2.10) to (2.15), as can be seen in Figure 3. This means that the Botzmann approximation used to derived the thermo-hydrodynamical equations was a good approximation.

3. PROBLEM OF THE DEFINITION OF THE TEMPERATURE

The temperature was introduced through the equipartition relation, which is satisfied for a classical gas. However, as already noticed, this leads to a particular temperature scale $1 \leq T \leq 2$. The basic reason for this is the fact that the particles of the lattice gas obeys a exclusion principle. Only one particle of a given velocity is present at a site at a given time. One

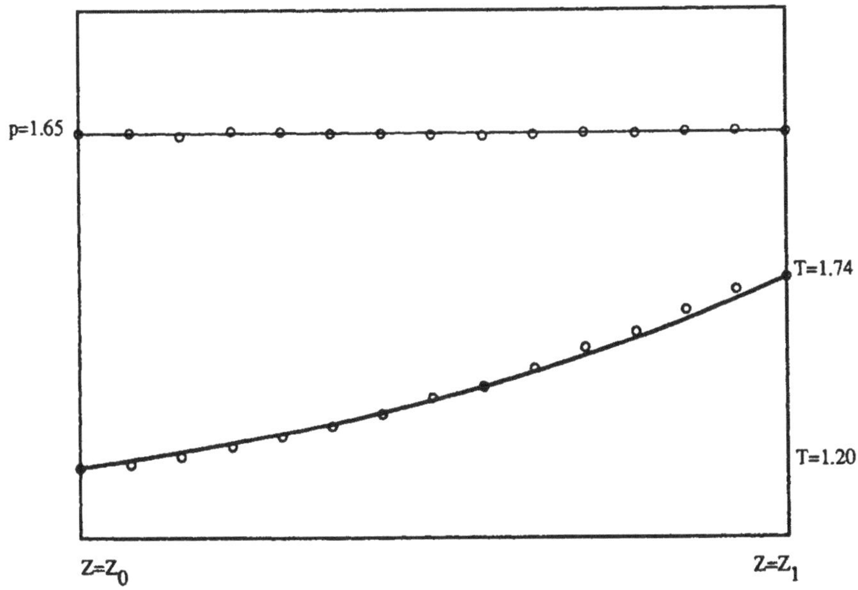

Fig. 3: Pressure and temperature profile between the plates. The circles represent the numerical data from the CA simulation. The solid lines are the theoretical predictions following the Boltzmann approximation of CA dynamics.

can then ask if there is a different way to introduce the temperature in this problem. An common definition of the temperature in equilibrium statistical mechanics is the microcanonical one, namely:

$$\frac{1}{T'} = k\left(\frac{\partial \ln W}{\partial E}\right)_V \tag{3.1}$$

where W is the number of allowed configurations, E is the total energy and V the volume that contains the particles. In our case, if the number of lattice sites is Ω, we have [6] :

$$W = \binom{4\Omega}{N}\binom{4\Omega}{M}$$

since, at each site, we can put up to four slow and fast particles.

We introduce the density as:

$$\rho = \frac{N+M}{\Omega}$$

using $k = \frac{v^2}{2}$, we have:

$$N = 2\rho\Omega - \frac{E}{k}$$

$$M = \frac{E}{k} - \Omega\rho$$

For Ω, N and M large, a straightforward calculation gives

$$\begin{aligned}\ln\Omega = 8\Omega\ln 4\Omega &- (4\Omega - 2\rho\Omega + \frac{E}{k})\ln(4\Omega - 2\rho\Omega + \frac{E}{k}) \\ &- (2\rho\Omega - \frac{E}{k})\ln(2\rho\Omega - \frac{E}{k}) - (\frac{E}{k} - \rho\Omega)\ln(\frac{E}{k} - \rho\Omega) \\ &- (4\Omega + \rho\Omega - \frac{E}{k})\ln(4\Omega + \rho\Omega - \frac{E}{k})\end{aligned}$$

Accordingly,

$$\frac{1}{T'} = \ln\frac{2-T}{T-1} + \ln\frac{\frac{4}{\rho} - (T-1)}{\frac{4}{\rho} - (2-T)} \tag{3.2}$$

where $T = \frac{E}{k\rho\Omega}$ is our first definition of the temperature. We recover the well-known results for a two level system. Namely, for a gas whose particles have all the same (slow) velocity $T' = 0$; for $M = N$, $T' = \pm\infty$ and T' is negative if $M > N$. Therefore this scale of temperature do not have the behaviour expected in a classical gas. The fermionic character of the lattice-gas is basically a artifact used to simplify the collision rules, but is not expected to show up at the macroscopic scale . A similar behaviour will still be present if ones were to allow a finite number of particules having the same velocity at one site at a give time. This is the reason why we have adopted a different definition of the temperature scale. Moreover, our choice gives a simple prescription to define a local nonequilibrium temperature $T(\vec{r})$, as the mean energy per particle in a small volume located at $\vec{r}$. Note that, due to this interpretation, it is simple to produce an external temperature gradient in the numerical simulation [5].

However, further work should be done to define a model in which the temperature scale has the desired macroscopic properties and is in agreement with the equilibrium statistical mechanics predictions.

4. EQUILIBRIUM AND NONEQUILIBRIUM CORRELATION FUNCTIONS

A more subtle problem is the study of the spatial correlation functions for the above two-dimensional fluid submitted to a temperature gradient.

The most common way of calculating correlation functions in a fluid is the generalised fluctuating hydrodynamics of Landau-Lifchitz [7] , where one adds random fluxes to the thermo-hydrodynamic equations. Unfortunately, even for simple nonequilibrium cases, one is confronted with great mathematical difficulties and one must appeal to drastic approximations. For this reason, numerical methods are often the most successfull approach [8] .

In the last part of this paper we shall show that cellular automata simulations can also be helpfull for computing these correlation functions . We shall restrict ourself to the density-density, temperature-temperature and density-temperature spatial correlation function in a stationary state.

To measure the correlation functions, we divided the area of the system into strips, parallel to the y-axes. The density of particle $\rho(z)$ in a strip at heigt z is obtained by counting the total number of particles into it. In the same way, the temperature $T(z)$ is given by the total kinetic energy, divided by the total number of particles of the strip. These two quantities are time dependent and, in a sationary state, they fluctuate around their mean values. The spatial correlation functions were obtained by averaging over time the products $\rho(z)\rho(z')$, $T(z)T(z')$ and $\rho(z)T(z')$, and then by substracting the respective products of the averages:

$$C_{\rho\rho}(z,z') \equiv < \rho(z,t)\rho(z',t) > - < \rho(z,t) >< \rho(z',t) > \tag{4.1}$$

$$C_{TT}(z,z') \equiv < T(z,t)T(z',t) > - < T(z,t) >< T(z',t) > \tag{4.2}$$

$$C_{\rho T}(z,z') \equiv < \rho(z,t)T(z',t) > - < \rho(z,t) >< T(z',t) > \tag{4.3}$$

The generic form of these correlations is

$$C = \frac{1}{V} f(z)\delta_{z,z'} - g(z,z') \tag{4.4}$$

where V is the area of the strip.

For the simulations, the strip at height z' was kept fixed at $z' = \frac{L}{2}$ and we measured the correlations for five values of z between 0 and L. For $z = z'$, the temperature-temperature and the density-density correlations both show a pronounced peak, whose height depends of the area of the strips. By varying the area of the strips we can calculate $g(z,z')$ which exibits the *long range correlations characteristic of a nonequilibrium* system [9] , [10] .

The functions $g(z, \frac{L}{2})$ are shown on Figure 4, for both an equilibrium (e.i. without the cold and the hot plates) and a nonequilibrium situation. The results of the numerical simulations are marked with a square for the

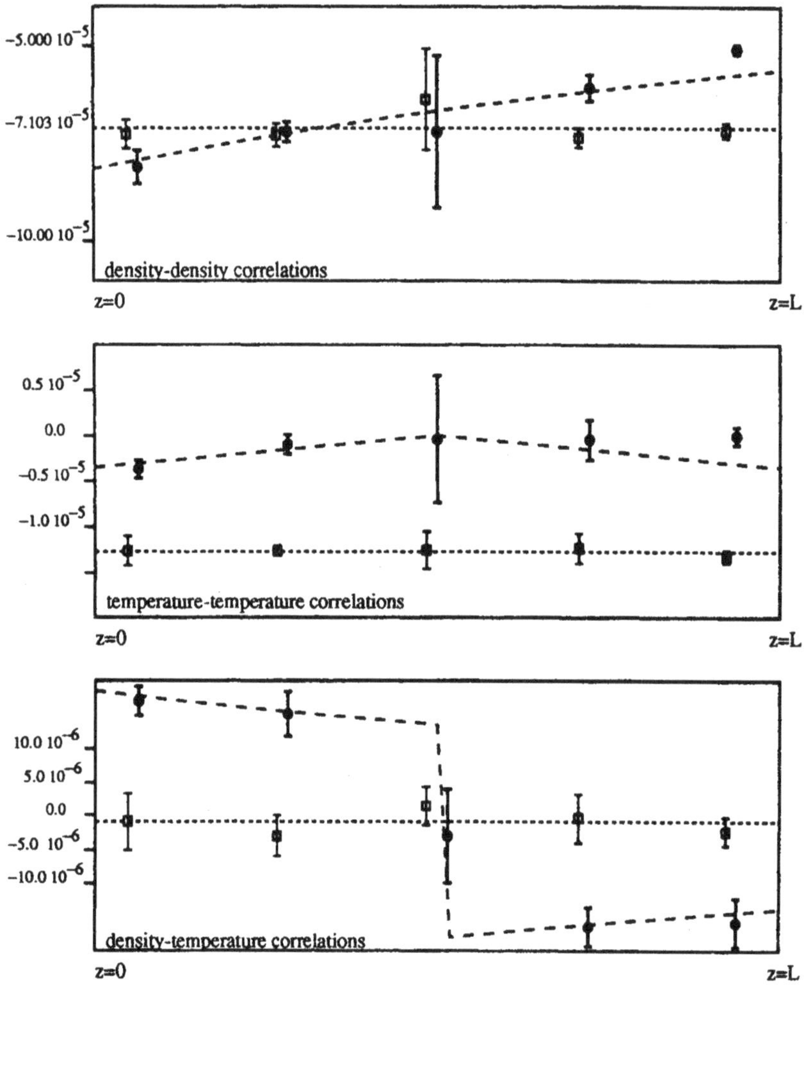

Fig. 4: Functions $g(z, L/2)$. The squares represent the experimental data for the equilibrium case while the dotted line is the prediction of the theory. The circles represent the experimental data for the nonequilibruim case. The dashed curves are the predictions from the fluctuating hydrodynamics.

equilibrium and with a circle for the nonequilibrium. These values can be compared with theoretical predictions. At equilibrium, a microcanonical hypothesis for the lattice-gas model allows to solve the problem exactly [6]. The functions $g(z, z')$ then express the finite size effects. The horizontal dot lines show the theoretical values of g.

For the nonequilibrium case, we plotted the prediction of the fluctuating hydrodynamics (dashed curves). Although these nonequilibrium calculations did not take into accout the particularity of the lattice-gas model, their predictions agree qualitatively well with the C.A. simulations.

In conclusion, we see that lattice-gas models provide an interesting alternative approach to describe, at least qualitatively, thermal effets in fluids. However, more general models are needed to describe thermal effets in fluids in motion and multi-speed FHP like models have been proposed [11] , [12]

Acknowledgments: Bastien Chopard is acknowledge for his close collaboration during the developpement of all the work exposed above.

REFERENCES:

[1] U. Frisch, D. d'Humières, B. Hasslacher, P. Lallemand, Y. Pomeau and J.-P. Rivet, in "Modern Approach to Large Nonlinear Systems", Complex Systems, 1, 649 (1987).

[2] D. d'Humières (in these proceedings)

[3] U. Frisch and J.P. Rivet C.R. Acad. Sci. II, **303**, 1065, (1986).

[4] L. Kadanoff Physics Today **39**, 7 (1986)

[5] B. Chopard and M. Droz, Phys. Lett. **A126**, 476 (1988)

[6] B. Chopard, PhD thesis, University of Geneva (1988)

[7] L. Landau and E. Lifchitz, "Statistical Physics", Pergamon (1981).

[8] M. Malek Mansour, A. L. Garcia, G. C. Lie and E. Clementi, Phys. Rev. Lett **58**, 874, (1987).

[9] J. M. Rubi in "Recent Developments in Nonequilibrium Thermodynamics", Lecture Notes in Physics, vol. **253**, J. Casas-Vázquez, D. Jou and J. M. Rubi eds. Springer 1985.

[10] R. Schmitz, Phys. Reports, **171**, 1, (1988).

[11] D. d'Humières and P. Lallemand, Europhys.Lett., 2(4), 291 (1986).

[12] B. Chopard and M. Droz, in "Chaos and Complexity", Torino 1987, R. Livi editor, World Scientific (1988), p 302.

NUMERICAL CALCULATION OF THE CONDUCTIVITY OF PERCOLATION CLUSTERS AND THE USE OF SPECIAL PURPOSE COMPUTERS

H.J. HERRMANN
Service de Physique Théorique†
de Saclay
91191 Gif-sur-Yvette Cedex, France

Abstract

Electrical conductivity diffusion or phonons, have an anomalous behaviour on percolation clusters at the percolation threshold due to the fractality of these clusters. The results that have been found numerically for this anomalous behaviour are reviewed. A special purpose computer built for this purpose is described and the evaluation of the data from this machine is discussed.

1. DYNAMICAL SCALING OF PERCOLATION

Percolation [1] has been a model for disordered systems, like porous media, random mixtures of conductors and insulators and gels. It is usually defined on a d–dimensional regular lattice on which sites (or bonds) are occupied with a probability p and empty with probability $1 - p$. Occupied nearest-neighbors belong to the same cluster. At a critical value p_c clusters of all sizes are present, particularly also arbitrarily large clusters. These large clusters are fractals, i.e., their density ρ decreases with their linear size L with a powerlaw $\rho \sim L^{d_f - d}$ where d_f is called the fractal dimension ($d_f = 91/48$ in $d = 2$ and $d_f \approx 2.5$ in $d = 3$).

A consequence of fractality is that the cluster is less connected on larger length scales, it is actually only singly connected on a number of bonds n_r that grows like $n_r \sim L^{d_r}$ ($d_r = 3/4$ in $d = 2$ and $d_r \approx 1.14$ in $d = 3$). Due to this unusual decrease in connectivity many fundamental physical laws are different on these fractal clusters from the corresponding law on an Euclidean lattice [2]. So, the well known diffusion law, $R \sim N^{1/2}$, where R is the average distance a diffusing particle has from its origin after N jumps, changes on a fractal to

$$R \sim N^{1/d_w} \tag{1}$$

which is called the "anomalous diffusion" law. The exponent d_w, the fractal dimension of the trace of the particle on the fractal cluster, is larger than 2 due to the weak connectivity mentioned above.

Another physical law that changes on a fractal is the size dependence of the electrical conductance σ. On an Euclidean lattice it goes like $\sigma \sim L^{d-2}$, on a fractal cluster like

$$\sigma \sim L^{-\zeta} \tag{2}$$

where

$$\zeta = 2 - d + \tilde{t}$$

and $\tilde{t} > 0$. This effect has actually been measured for instance for Pb on thin Ge and Au films [3]. Electrical conductivity and diffusion are related to each other via the Einstein relation and one can show [4] that

$$\zeta = d_w - d_f \tag{3}$$

The phonon spectrum $\rho(\omega)$ usually behaves for small frequencies ω like $\rho(\omega) \sim \omega^{d-1}$, on a fractal one has instead $\rho(\omega) \sim \omega^{d_s - 1}$, where the spectral dimension d_s is also related to the previous exponents [5] via $d_s = 2d_f/d_w$.

Intrigued by the question if it is possible to relate this critical exponent to usual static critical exponents, particularly to d_f, several conjectures have been proposed. The most serious ones were

$$d_w = d_f + 1 \tag{4a}$$

(proposed first for $d = 2$ in Ref [6]),

$$d_w = \frac{3}{2} d_f \ [5] \tag{4b}$$

(called the Alexander-Orbach conjecture) and for $d = 2$ only

$$d_w = 2(d_f - 1 + d_f^{-1}) \ [7] \tag{4c}$$

For $d = 2$ they give in the same order $\tilde{t}$ =1, 0.948 and 0.951. Finally it was also proposed that $d_w = d_f + 1$ for $d_f \leq 2$ and $d_w = \frac{3}{2} d_f$ for $d_f > 2$ [8]. It became clear very soon that none of these relations has a validity for arbitrary fractals as was thought for a moment since they clearly do not work for Sierpinski gaskets or lattice animals [2]. Neither is any of these conjectures able to reproduce the $\varepsilon = 6 - d$ expansion [9] even in lowest order in ε. On the other hand new arguments favouring some of these conjectures came up [10].

2. NUMERICAL METHODS

To shed light upon the question raised by the various conjectures it is necessary to make numerical high-pression calculations. This can be done via diffusion or via the electrical conductivity. The largest effort done by averaging over many diffusion trajectories gave in $d = 2$, $\tilde{t}$ =0.960±0.015 [11] which is not very conclusive. More efficient was the evaluation of the probability cloud, i.e., on each site the value of the probability that the diffusing particle is there after N time steps provided it was at the beginning on the origin with probability one. This method automatically averages over all trajectories and $\tilde{t}$ =0.97±0.01 [12] was obtained from graphs as shown in Fig.1.

The calculation of the electrical conductivity of percolation clusters can also be done in different ways. Using star-triangle transformations it is possible to reduce the conductance of any $2d$ network to that of one bond [13] and this method, pushed to systems of size 1000, yields [13] $\tilde{t} = 0.977^{+0.003}_{-0.010}$. The other way to calculate the conductance is adding bond after bond to the cluster; to do so one has to use a strip geometry.

This method, often called strip method, was introduced in Ref. [14]. Top and bottom of the strip have a fixed potential (fixed boundary condition in the longitudinal direction). If the strip has width n one defines by I_i and V_i the current and potential drops at the end of each of the n lines and the conductance matrix σ_{ij}, given through

$$I_i = \sum_{i=1}^{n} \sigma_{ij} V_j \tag{5}$$

is updated via

$$\sigma'_{ij} = \sigma_{ij} - \sigma_{i\alpha}\sigma_{\alpha j} r/(1 + \sigma_{\alpha\alpha} r) \tag{6a}$$

if a longitudinal bond is added at line α and via

$$\sigma'_{ij} = \sigma_{ij} + (\delta_{\alpha j} - \delta_{\beta j}) \times (\delta_{\alpha i} - \delta_{\beta i})/r \tag{6b}$$

if a transversal bond is added between lines α and β where r is the resistance of the bond. The statistical fluctuations become very small for long strips and the extrapolation to infinite width is just given by

$$\sigma_{11} \sim Ln^{-\tilde{t}} \tag{7}$$

This method gave [15] $\tilde{t} = 0.973 \pm 0.005$ in $2d$.

Instead of having a conductor on each occupied bond and an insulator on each empty bond one can put a superconductor ($r = 0$) on each occupied bond and a normal conductor on each empty bond. In this new problem also a new, independent exponent, $\tilde{s}$, appears.

It happens, however, that in $d = 2$ one has $\tilde{s} = \tilde{t}$ from duality [16]. Calculating this superconductor problem via the strip method [17] has the advantage over the normal conductor problem of the previous paragraph that one can have periodic (instead of fixed) boundary conditions in the longitudinal direction which considerably reduces the finite size effects. One considers now the resistance matrix R_{ij}, defined through

$$V_i = \sum_{i=1}^{n} R_{ij} I_j \tag{8}$$

and updates this matrix via

$$R'_{ij} = R_{ij} - r\delta_{i\alpha}\delta_{\alpha j} \tag{9}$$

if a longitudinal bond at line α is added and via

$$R'_{ij} = R_{ij} - (R_{i\alpha} - R_{i\beta})(R_{\alpha j} - R_{\beta j})/(r + R_{\alpha\alpha} + R_{\beta\beta} - R_{\alpha\beta} - R_{\beta\alpha}) \tag{10}$$

if a transversal bond between lines α and β is added. For large widths one has

$$R_{11} \sim Ln^{-\tilde{s}} \tag{11}$$

and

$$\tilde{t} = 0.977 \pm 0.010 \tag{12}$$

was obtained with a moderate effort ($\approx$ 5h Cray 1) [17] as seen in Fig. 2.

In $d = 3$ the up to now most precise values for the exponents were obtained with the strip method and they are $\tilde{t}$ =2.2±0.1 [18] and $\tilde{s}$ =0.85±0.05 [17]. They exclude most of the conjectures but not the Alexander-Orbach conjecture. We note however, that recent [19] diffusion simulations found $\tilde{t}$ =2.8±0.3 in disagreement with previous work.

3. BUILDING A SPECIAL PURPOSE COMPUTER

In order to get the exponents with a significantly higher precision much more computer power has to be inverted. For this purpose the most efficient method, namely that of Ref [17] was chosen and a special purpose computer (SPC) was build to implement this method in hardware.

Isn't it much more comfortable to write a Fortran program and execute it on the next available general supercomputer (like a Cray XMP, an IBM 3090, a Cyber 205, etc...)? Several reasons have led people to do it the hard way:

First of all, general computers must usually be shared with many other users, so one will have access to a restricted number of hours, several hundred hours per year at best. An SPC is available every day of the year, 24 hours a day (except during blackouts).

There also is an important financial aspect: A grant proposal asking for 2×10^7 US dollars for a Cray 2 is likely to be rejected. The chances, however, to get 5×10^4 US dollars for tools and components to build an SPC are considerably better.

Finally there is speed! Supercomputers are famous for being very fast (that is what makes them price-efficient compared to smaller general computers). Now, however, there exist on the market highly integrated ultrafast chips. If, in addition, one arranges these chips inside the SPC in such a way that many of them work simultaneously (doing the same thing (parallelism), or doing different things (vectorization)) the SPC can be as fast or even faster than a supercomputer. Technologically speaking, this is the reason why SPC's are mainly a development of only the last couple of years.

All these reasons in favour of SPC's have to be counterbalanced by the fact that the building of an SPC involves an enormous amount of work in a field that usually has to be learned from scratch.

Since the building of an SPC is a non-negligible enterprise one should first consider carefully if one has chosen the right problem. There are certain general criteria:

- The problem one wants to solve should be sufficiently fundamental and sufficiently difficult that after two years - which is about the average time one needs to get results from an SPC - the scientific community is still interested.

- The algorithm one uses must be relatively simple, so that the SPC won't get too complex and its construction time too long.

- The method to solve the problem should involve a huge repetition of identical steps (typically 10^8 to 10^{10}) so that one can run the SPC for many days without intervening.

The algorithm and the evaluation of the data should have had extensive previous testing on general computers and one should be so sure about the procedure that one is willing to fix in a definitive way what calculations one will perform. It is usually very difficult to change an SPC once it is built.

Once one has chosen the algorithm for which one wants to build the SPC there are certain necessary conditions required before construction can start: One needs the financial support to buy the hardware components and electronic equipment (like an oscilloscope or a logic analyzer). An enthusiastic team willing to work hard must be formed. At least one person with a solid experience in electronics must belong to the team. Finally one has to find an electronics lab willing to put a reasonable infrastructure at your disposal.

4. THE SPC IN SACLAY

In the following I will briefly describe some aspects of the construction of the SPC that was build in Saclay [20] by M. Hajjar and J.M. Normand to calculate the electrical conductivity of percolation clusters using the strip method.

The first, and most creative step is the design of the architecture of the SPC. One has to try to decompose the algorithm into the largest possible number of elementary calculations that can be performed simultaneously. Some calculations are completely independent from the rest, like the generation of random numbers, and can therefore be executed by a processor of its own (MIMD = multiple instruction multiple data arrangement), so they are implemented on a separate electronic board.

For the part of the calculation that will be executed the most one has to construct a "pipeline". In our case the main calculation is the updating of a matrix R_{ij} following eq. (10):

$$R'_{ij} = R_{ij} - R_{11} - V_i * V_j \ , \tag{13}$$

where V_i is a vector and all operations are in 64 bit floating point arithmetic. In fig.3 we see a simplified diagram of the data flow. Like on a car assembly line the first subtraction of eq. (13) is performed for one matrix element while the second subtraction of eq. (13) is done for another matrix element (MISD = multiple instruction single data arrangement). In order that such a pipeline can work every unit must be finished at the same time. So one has to investigate which chips are available on the market. It happens that the fastest 64 bit multipliers need 480 nsec per multiplication (WTL 1064 of Weitek) while a subtraction can be done in 120 nsec (WTL 1065). So one has to put four multipliers in parallel as shown in fig. 3 (SIMD = single instruction multiple data arrangement) and one can get a result every 120 nsec.

The design of the architecture is, however, far from being finished! In fig. 3 R_{ij} and V_i are stored in memories. One needs a device, the address generator, that determines the location of a given matrix element within the memory. Then one needs somewhere to store the "vector-instructions", i.e., the sets of bits that tell each of the chips what function it has to perform. The ensemble of all these vector-instructions is called the microcode. Which instruction is to be applied and how often is managed by a "sequencer". Roughly speaking the sequencer makes the loops and jumps that we are accustomed to use in Fortran programs. All these units must have the same speed as the central processing unit (CPU) of fig. 3, i.e., 120nsec.

Finally a connection to the external world must be established. One needs a host computer, an in-house Motorola 68000 based card in our case, that will receive the data coming from the CPU, evaluate it and store it on a disk. The host computer need not be fast but must be reliable. From time to time it will make a complete backup of the state of the machine and make consistency checks: It is rather complicated to use error correcting codes as those implemented on the big general computers, so if one finds an error during a calculation (due for example to a cosmic ray) it is restarted using the last backup as a starting point.

The host computer will not work with 64 bits and will, since it is slow, not be synchronous to the CPU, so an interface between the two is needed which is built on a separate board. The whole ensemble is sketched in fig. 4.

After the architecture is settled important electronic decisions must be taken: Which technology should be used to build the electronic boards: wirewrap or printed circuits, or should one even make integrated circuits ? In Saclay, we opted for printed boards in TTL logic. Also one has to choose all the chips out of the data books of the different manufacturers and to compare them one has to carefully study their specifications. Since about 200 different chips were needed this is a major effort.

After all these decisions are made, a detailed layout has to be made for each board. Each pin of each chip must be properly connected and, of course, the chips must fit on the board.

In order that the connections are not too long and do not cross each other too much the placing of the chips on the board and the routing of the connection lines between them has to be done very carefully. CAD (= computer assisted design) programs are commercially available and can be extremely useful for this purpose. CAD programs can write all information concerning the locations of chips and wires on tape. These tapes can be sent to a company that will make the boards for you within several weeks.

In the meantime one orders the chips from the respective manufacturers and builds the backplane into which the boards can be placed. Everything is put together and the SPC is ready for testing. The debugging will take some more months before actual calculations can start.

5. RESULTS OBTAINED ON THE SPECIAL PURPOSE COMPUTER IN SACLAY[21]

The machine is microprogrammable and turns at 25 Mflops so that for our type of calculation it is about 10 % faster than the vectorized program run on one processor of the Cray X-MP. Memory allows to calculate strips of widths $n \leq 255$ in two dimensions. Technical details about the machine are given elsewhere [20].

All calculations were done on the square lattice. We considered bond percolation at $p_c = 0.5$ and site percolation at $p_c = 0.592745$ (2) [22]. The strips are at least 100 times longer and the statistical error bars at least 10 times smaller than in ref. [17]. One hundered days of CPU time were needed to obtain these data, fifty days alone for the two values for $n = 40$ since the time required grows with n^3.

While generating the above data we encountered an unexpected problem that turned out to be due to the random numbers. Random numbers X_k are generated through the lagged-Fibonacci sequence

$$X_k = (X_{k-r} + X_{k-s}) \bmod 2^{32} \tag{14}$$

Using the generator $(r, s) = (17, 5)$ of (4), which had been implemented also on an other special purpose computer [32], we observed for bond percolation at $n = 9$ a spurious deviation of the data from the expected monotonic behaviour by about ten times the size of the error bars (independent of the seed). We also encountered weaker, significant deviations at some other sizes (e.g. $n = 5$) and also for the generators $(r, s) = (31, 13)$ at

$n = 10$ and $(r, s) = (55, 24)$ at $n = 20$. Although we do not understand the origin of these flaws they are apparently inherent to the family of generators of (14). These problems could only be observed because of the long sequences (up to 10^{11}) that we generated, still the periods of the three generators are respectively larger than 10^{14}, 10^{18} and 10^{25}. We note that in other cases where sequences of this length have been used similar problems were observed [28]. Because of this problem we were forced to repeat our calculation for a large part of the values with up to three different generators in order to identify the biased data. So, effectively we spent much more CPU time.

In fig. 5, we show our data multiplied by $n^{0.98}$ as a function of n in a log-log plot. Due to universality the curves should become parallel, straight lines of slope $0.98 - \tilde{t}$ for large n. Because of the corrections to scaling they are curved, fortunately in the opposite way. So, the bond percolation data imply $\tilde{t} \leq 0.98$ while the site percolation data imply $0.96 \leq \tilde{t}$ excluding therefore already both conjectures 91/96 of eq. (4b) and 1 of eq. (4a).

We tried several fits based on the minimalization of the square distance from the individual data points. A logarithmic prefactor fits very badly and can be excluded. Fitting with one powerlaw correction gives leading exponents for site and bond percolation that differ by at least one percent. Fitting a logarithmic correction gives about the same result. The precision of our data should, however, allow for a determination of $\tilde{t}$ better than one percent.

To better understand the above problem we explicitly imposed that in the relation

$$R_n = n^{-\tilde{t}}(c_1 + c_2 n^{-\omega}) \tag{15}$$

not only $\tilde{t}$ but also the leading correction exponent ω should be universal (c_1 and c_2 are non-universal constants). So we plotted $R_n\, n^{\tilde{t}}$ against $n^{-\omega}$ for different choices of $\tilde{t}$ and ω and tried to find $\tilde{t}$ and ω such that both site and bond percolation data fall on straight lines. This is not possible reflecting the problem found before. For site percolation a straight line can be obtained for reasonable values of ω (≈ 1.2). For bond percolation the data are never aligned and for the parameters for which site percolation data are straight the bond percolation data line on an S shaped curve. In fig. 6 we show such a situation. It becomes therefore clear that higher order corrections play an important role in bond percolation. Only for sizes $n \geq 7$ the leading correction seems sufficient since the data seem to follow a straight asymptotic line (see fig. 6).

We explored for which exponents the data are reasonably consistent with the scenario of fig. 6, i.e. straight lines for all site percolation data and for bond percolation with $n \geq 7$. We found

$$\tilde{t} = 0.9745 \pm 0.0015 \quad \text{and} \quad \omega = 1.2 \pm 0.2 \tag{16}$$

The value of $\tilde{t}$ falls into the error bars of all previous estimates [12,15,17] and its error bar is several times smaller than previous error bars.

Besides the very precise calculation of the dynamical exponent of $2d$ percolation, which excludes the two conjectures 91/96 and 1, we estimated the leading correction to scaling exponent and found that in the case of bond percolation sizes $n < 7$ are dominated by even higher corrections. We discovered that random number generation is the most critical numerical difficulty and we believe that for the future of high speed computation it is of fundamental importance to find generators that have no pathologies when one has sequences of the order 10^{11}. Our calculation would not have been possible without the special purpose computer and work calculating $\tilde{t}$ and $\tilde{s}$ in three dimensions is on course.

6. A SHORT REVIEW OVER SOME SPECIAL PURPOSE COMPUTERS THAT HAVE BEEN BUILT FOR STATISTICAL PHYSICS

Most of the about twenty SPC's that have been built by physicists in the last five years are related to statistical physics. Their construction time ranges from four months to four years, their price from 200 US dollars to 2×10^5 US dollars. About half of the SPC's built are restricted to very specific applications; the others are more general. In addition there are at least 30 projects for SPC's about which I will not talk. A partial list of the more specific existing SPC's that are related to statistical physics is given in table I.

Table I

Some SPC's built for statistical physics.

Place	Type	Completed	Ref.
M.I.T., USA	cellular automata	1982	23
E.N.S., France	cellular automata	1986	24
Santa Barbara, USA	Ising	1982	26
Delft, Netherlands	Ising	1982	30
Bell Labs, USA	Ising (spin glass)	1984	32
Delft, Netherlands	molecular dynamics	1983	36
Saclay, France	random conductivity	1987	20

Cellular automata have been the cheapest [23] and the shortest times were needed to construct them [24]. Since cellular automata SPC's can visualize their configurations very nicely on a screen, many interesting pictures have been made with them. Quantitative results have, however, been quite rare up to now. This might change in the future if recent ideas by Pomeau and others [25] to model hydrodynamic flow through cellular automata are exploited.

SPC's of the Ising type have been more successful in yielding precision data. The Santa Barbara machine [26] held the world record for a while for the speed of simulation of an Ising model with 30 million updates per second. (This record was later broken again by general computers [27]). It yielded precise values for the critical temperature and the dynamic critical exponent of the three dimensional Ising model making nearly 10^{12} updates [28]. However, more or less at the same time about equally precise estimates were obtained by other methods [29]. Also some incorrect results were produced for certain system sizes which were presumably due to correlations in the random number generator as was suggested for instance by calculations on the DISP [30]. The DISP is an SPC, slower but more versatile than the Santa Barbara machine. Several other calculations have already been performed on it including some that were reported on at this conference [31].

Another successful Ising-type SPC is the spin glass machine [32] built in Bell Labs. They found evidence for a finite transition temperature in the three-dimensional short range Ising spin glass [33]. The same result was also found by other authors at the same time using conventional computers [34]. The machine at Bell Labs was also used to study the random field Ising model [35].

Much more complex than the machines described above are molecular dynamics SPC's since they involve real numbers instead of integers and since they involve long range instead of short range interactions. The first molecular dynamics SPC was built by Bakker [36] in a pioneering effort using fixed point arithmetic. This machine was used to give some evidence that in two-dimensional melting no hexatic phase exists [37] in agreement with work done on conventional computers [38]. Another SPC for molecular dynamics has just been completed in San Jose [38].

Finally a number crunching SPC of equal order of difficulty is the one built in Saclay [20]. Its performance and the results obtained with it have been presented above.

Summarizing, we have described a new effort that has been made in the field of computational physics to overcome usual limitations of general supercomputers. It has to be left to the future to see if this development will survive in the long run or if a new generation of supercomputers will render it obsolete. Some more information about this field can be retrieved from older review articles [39].

REFERENCES

[1] Stauffer, D., "Introduction to Percolation Theory", Taylor and Francis, London (1985).

[2] Havlin, S. and Ben-Avraham, D., Adv. in Physics 36, 695 (1987).

[3] Kapitulnik, A. and Deutscher, G., Phys. Rev. Lett. 49, 1441 (1982).

[4] Gefen, Y., Aharony, A. and Alexander, S., Phys. Rev. Lett. 50, 77 (1983).

[5] Alexander, S. and Orbach, R., J. Physique Lett., 43, L625 (1982).

[6] Levinshtein, M.E., Shur, M.G. and Efros, A.L., Zh. Eksp. Teor. Fiz. 69, 2203 (1975).

[7] Coniglio, A. and Stanley, H.E., Phys. Rev. Lett. 52, 1068 (1984).

[8] Aharony, A. and Stauffer, D., Phys. Rev. Lett. 52, 2368 (1984).

[9] Wang, J. and Lubensky, T.C., Phys. Rev. B33, 4998 (1986).

[10] Aharony, A., Alexander, S., Entin-Wohlmann, O. and Orbach, R., Phys. Rev. Lett. 58, 132 (1987) and Aharony, A., private communication.

[11] Pandey, R.B., Stauffer, D., Margolina, A. and Zabolitzky, J.G., J. Stat. Phys. 34, 427 (1986).

[12] Hong, D.C., Havlin, S., Herrmann, H.J. and Stanley, H.E., Phys. Rev. B30, 4083 (1984).

[13] Frank, D.J. and Lobb, C., Phys. Rev. B37, 302 (1988).

[14] Derrida, B. and Vannimenus, J., J.Phys. A15, L557 (1982).

[15] Zabolitsky, J.G., Phys. Rev. B30, 4077 (1984).

[16] Dykhne, A.M., Zh. Eksp. Teor. Fiz. 59, 110 (1970).

[17] Herrmann, H.J., Derrida, B. and Vannimenus, J., Phys. Rev. B30, 4080 (1984).

[18] Derrida, B., Stauffer, D., Herrmann, H.J. and Vannimenus, J., J. Physique Lett. 44, L701 (1983).

[19] Pandey, R.B., Stauffer, D. and Zabolitzky, J.G., J. Stat. Phys. 49, 849 (1987).

[20] Hayot, F., Herrmann, H.J., Normand, J.M., Farthouat, P. and Mur, M., J. Comp. Phys. 64, 380 (1986); Hajjar, M., Thesis (1988); Normand, J.M. and Hajjar, M., IEEE Trans. Comp. and Normand, J.M., ACM Conference Proceedings St Malo 56 (1988).

[21] Normand, J.M., Herrmann, H.J. and Hajjar, M., J. Stat. Phys. 52, 441 (1988).

[22] Ziff, R.M. and Sapoval, B., J. Phys. A19, L1169 (1986).

[23] Toffoli, T., Physica 10D, 195 (1984).

[24] d'Humières, D., private communication.

[25] Pomeau, Y., J. Phys. A17, L415 (1984); Frisch, U., Hasslacher, B. and Pomeau, Y., Phys. Rev. Lett. 56, 1505 (1986); Margolus, N., Toffoli, T. and Vichniac, G., Phys. Rev. Lett. 56, 1694 (1986).

[26] Pearson, R.B., Richardson, J.L. and Toussaint, D., J. Comp. Phys. 51, 241 (1983).

[27] Reddaway, S.F., Scott, D.M. and Smith, K.A., Comp. Phys. Comm. 37, 351 (1985); Herrmann, H.J., J. Stat. Phys. 45, 145 (1986); Okabe, Y. and Kikuchi, M., Statphys. 16 conference.

[28] Barber, M.N., Pearson, R.B., Toussaint, D. and Richardson, J.L., Phys. Rev. B32, 1720 (1985); Pearson, R.B., Richardson, J.L. and Toussaint, D., Phys. Rev. B31, 4472 (1985).

[29] Pawley, G., Swendsen, R.H., Wallace, D.J. and Wilson, K.G., Phys. Rev. B29, 4030 (1984); Adler, J., J. Phys. A16, 3565 (1983); Kalle, C., J. Phys. A17, L801 (1985); Wansleben, S. and Landau, D.P., Statphys. 16 conference.

[30] Hoogland, A., Spaa, J., Selman, B. and Compagner, A., J. Comp. Phys. 51, 250 (1983).

[31] Blöte, H.W.J., Compagner, A. and Hoogland, A., Physica A, 141, 375 (1987); Blöte, H.W.J., Compagner, A., Cornelissen, P.A.M., Hoogland, A., Mallezie, F. and Vanderzande, C., Physica 139A, 395 (1986).

[32] Condon, J.H. and Ogielski, A.T., Rev. Sci. Instr. 56, 1691 (1985).

[33] Ogielski, A.T. and Morgenstern, I., Phys. Rev. Lett. 54, 928 (1985); Ogielski, A.T., Phys. Rev. B32, 7384 (1985).

[34] Bhatt, R.N. and Young, A.P., Phys. Rev. Lett. 54, 924 (1985); see also Singh, R.R.P. and Chakravarty, S., Phys. Rev. Lett. 57, 245 (1986); Reger, J.D. and Zippelius, A., Phys. Rev. Lett. 57, 3225 (1986).

[35] Ogielski, A.T. and Huse, D.A., Phys. Rev. Lett. 56, 1298 (1986).

[36] Bakker, A.F., Bruin, C., van Dieren, F. and Hilhorst, H.J., Phys. Lett. A93, 67 (1982).

[37] Bakker, A.F., Bruin, C. and Hilhorst, H.J., Phys. Rev. Lett. 52, 449 (1984).

[38] Abraham, F.F., Adv. in Phys. 35, 1 (1986).

[39] Hilhorst, H.J., Bakker, A.F., Bruin, C., Compagner, A. and Hoogland, A., J. Stat. Phys. 34, 987 (1984); Hilhorst, H.J. and Herrmann, H.J., J. Phys. Lett. B42, 52 (1986); Science 215 (1982) February issue; Phys. Today 36 (1983) May issue.

[40] Herrmann, H.J., Physica A140, 421 (1986).

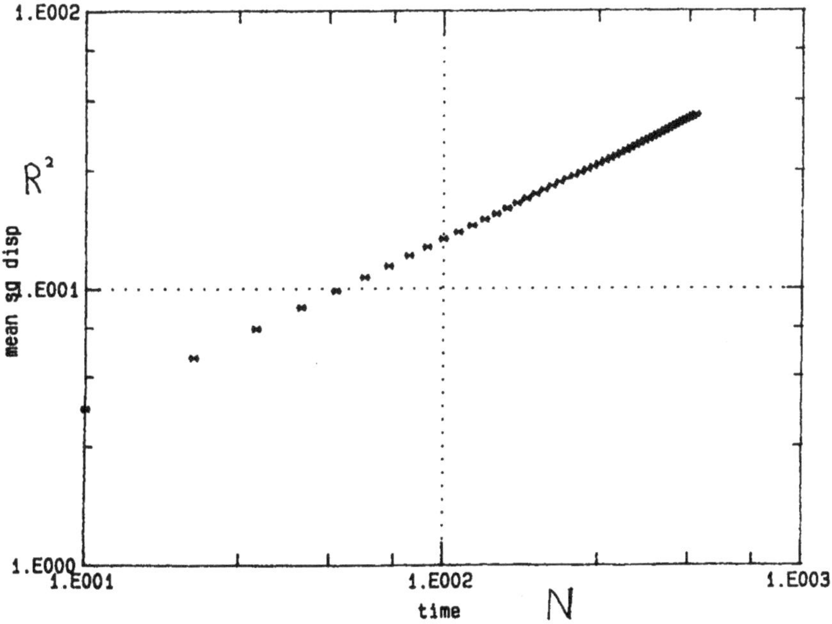

Fig. 1 - log-log plot of the square distance R^2 as a function of time steps N for a system 40×40 at p_c averaging over 200 samples. The slope is $2/d_w$.

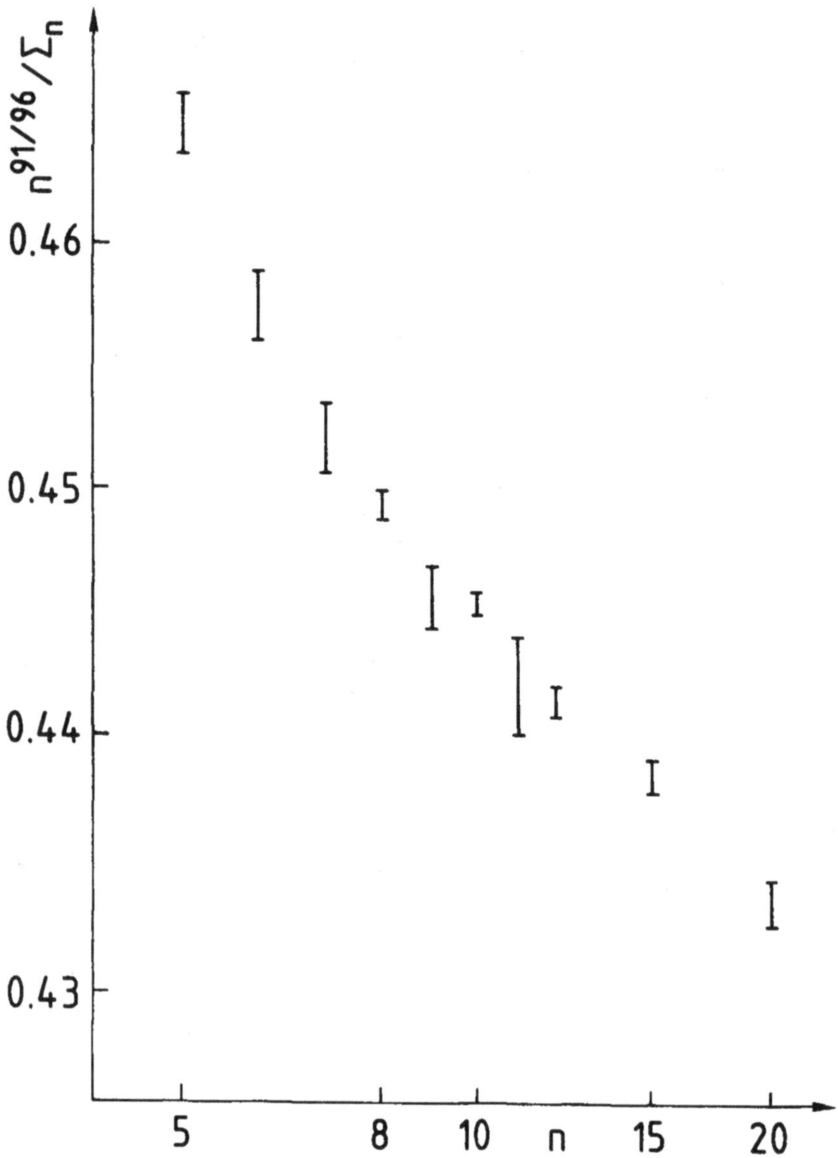

Fig. 2 - log-log plot of $n^{91/96}/\Sigma_n$ against n for $2d$ bond percolation. The error bars are statistical. If conjecture (4b) were correct one would have asymptotically a horizontal line. Taken from Ref. 17.

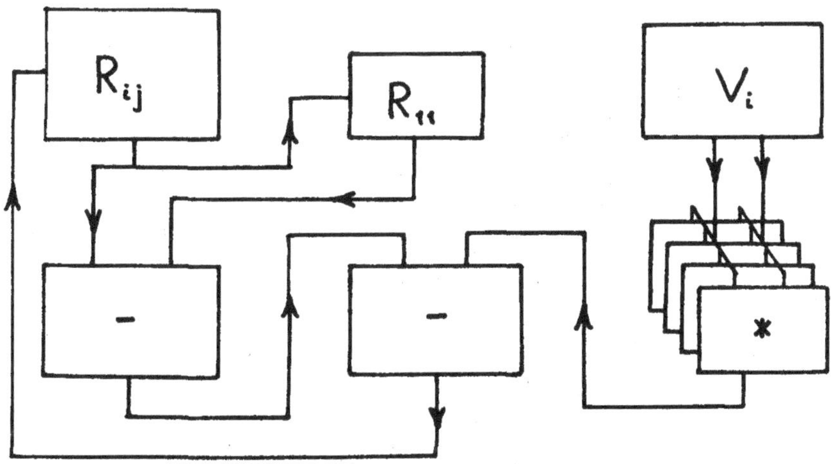

Fig. 3 - Schematic diagram for an architecture to calculate eq. (13). Taken from Ref. 40.

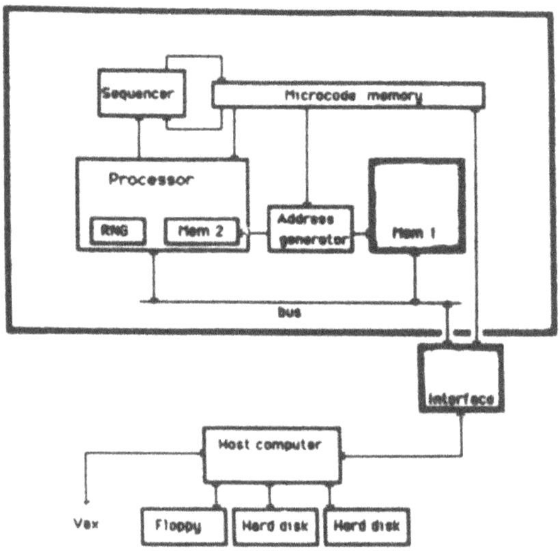

Fig. 4 - General diagram of the Saclay machine [20]. The parts surrounded by heavy lines are the ones actually built. Taken from Ref. 40.

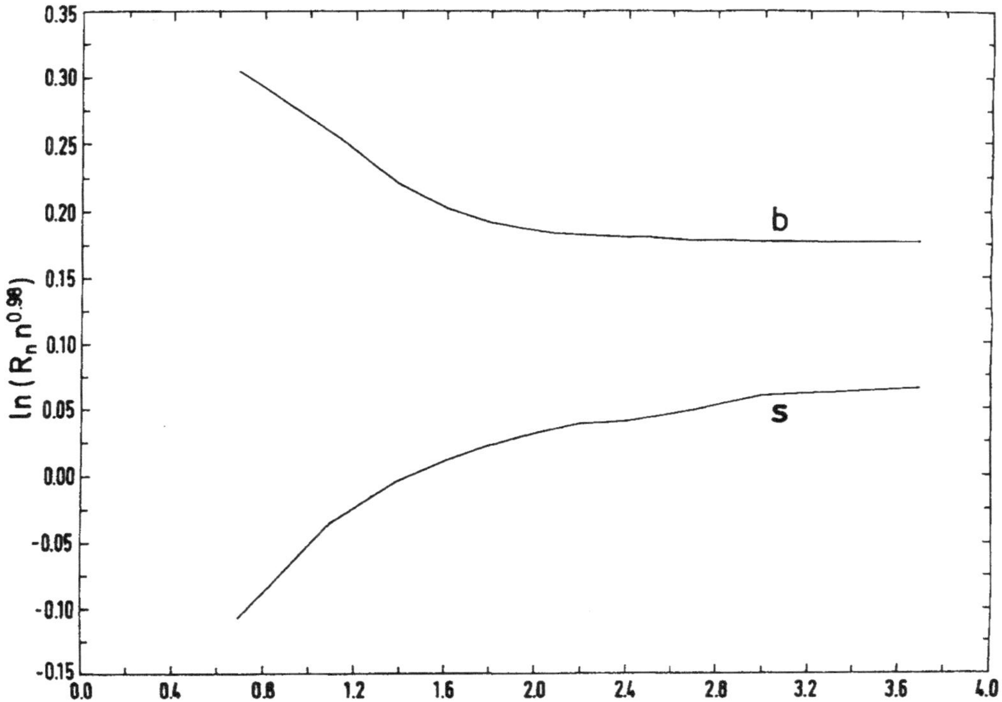

Fig. 5 - log-log plot of $R_n n^{0.98}$ against n for bond (b) and site (s) percolations. The bond percolation date are multiplied by a factor 2.5 to allow showing all data in one figure. Taken from Ref. 21.

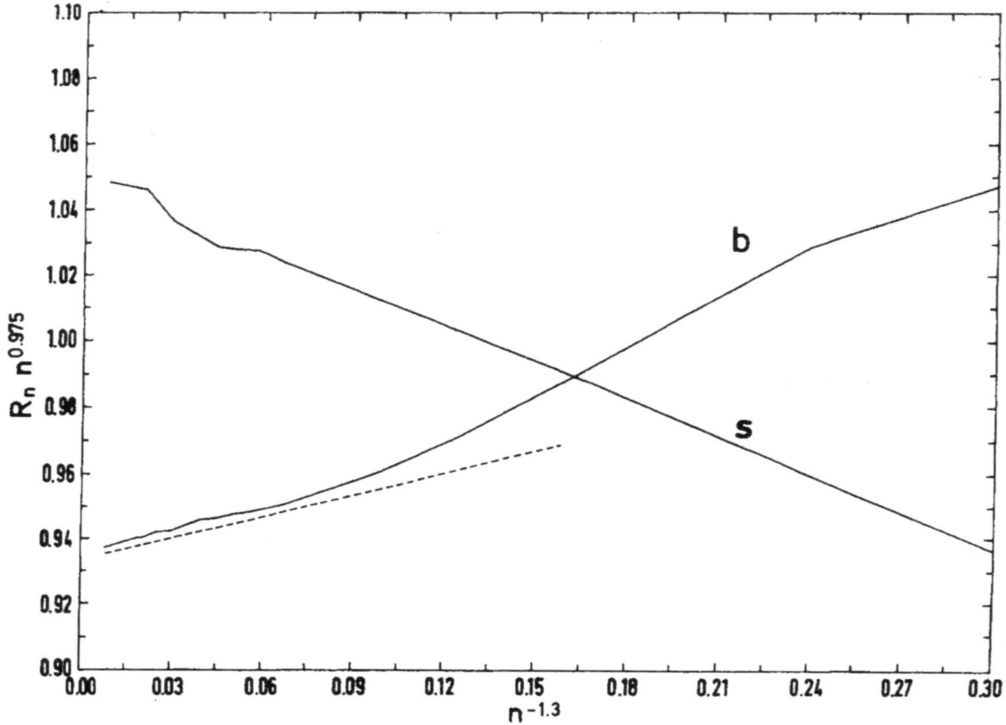

Fig. 6 - $R_n n^{0.975}$ against $n^{-1.3}$ for bond (b) and site (s) percolations. The bond percolation data are multiplied by two to allow showing all data in the same figure. Taken from Ref. 21.

IMMUNOLOGICALLY MOTIVATED CELLULAR AUTOMATA

D.Stauffer, HLRZ, c/o KFA, 5170 Jülich 1, West Germany

Cellular automata are an old field of research and have found several applications, as is seen from other talks at this conference. This mini-review, based mainly on 1989 work of Neumann, Pandey, and Chowdhury, uses cellular automata in an attempt to describe immunology, both in healthy and in sick individuals.

Reading papers by real (i.e. experimental) immunologists (e.g. by Cohen[1]) one sees a wide difference between their style and that of the cellular automata or Ising community. Does it make sense to describe the complications of life by such simple models like spins being up or down only ? Of course, reality is different and one can easily prove each of the models below to be unrealistic. However, studying complex reality alone does not guarantee success, as can be seen from cancer research. And sometimes success in science came from unrealistic models. For example, Kepler's laws approximated the earth by a point, even though everybody knew this to be wrong. And the Ising model ignores the well known fact that magnets can be bent and put into vibrations. Nevertheless, Kepler's laws started theoretical physics, and the Ising model is still a standard research subject of statistical physics. Of course, one should not expect Kepler's laws to describe continental drift, or the Ising model to describe the phonon spectrum of iron.

The real problem of these immunology models is that after only a dozen papers it is unlikely that the best model, corresponding to Kepler's laws or the Ising magnet, have already been found. Thus presumably everything here is wrong, except some methods. But even then there is some faint hope: Perhaps these models instead describe something else, just as the Ising model describes liquid-gas critical points. Columbus did not find what he was looking for, a sea lane to India; nevertheless his travels had some impact.

A healthy immune system consists of B cells (bone-marrow derived lymphocytes), producing (indirectly) antibodies which attack and neutralize foreign antigens like some virus. They are helped by helper cells and suppressed by suppressor cells. Macrophages eat the antigens and present parts of that food on their cell surface in order to alarm helper cells. Infection by some virus can stimulate the immune system in such a way that a second infection is dealt with much better. Vaccination against smallpox made so many people immune that this disease seems eradicated since several years. A more recent application of immunological knowledge is T cell vaccination[1]. The immune system can go wrong by either attacking the own body (auto-immune diseases like multiple sclerosis or diabetes mellitus), or by being weakened through the AIDS virus. These three cases, the normal, the too strong, and the too weak immune response, will be treated here; see our table.

Atlan[2] outlined the general philosophy of this cellular automata approach: We assume the concentrations of each cell type to be either very high (spin up, 1, true), or to be very low (spin down, 0, false). We assume that these cell variables interact by some deterministic rules, which we express as Boolean relations. Each cell variable at time t+1 is determined by itself and the other cell variables at the previous time step. In this way one has a simple mean field theory, the whole human body being invaded by a homogeneous concentration of cells, with one variable only for each cell type. (In response to one specific a antigen, we deal only with the part of the immune system reacting to that antigen, not the other antibodies etc.)

Such a mean field approximation of spatially homogenous concentrations is presumably good in most cases since fluids spread through the body in minutes and hours, whereas the immune response takes days. In special cases, one might also have a slonwly growing cancer-like immune response, which could be modelled on a lattice. Each lattice site now carries one set of spin variables, as used in the mean field model, and spins on different neighboring lattice sites interact.

NORMAL IMMUNE RESPONSE

The extended KUT model[3,4], for the normal immune response deals with antibodies A, suppressors S, helpers H, B-cells B, and the external antigen V (for virus; denoted as E in ref.3). Each of these five spins A, S, H, B, and V can be up (true, concentration 1) or down (false, concentration 0). The dynamics is governed by

A = V.and.B.and.H

S = H.or.S

H = H.or.(V.and..not.S)

B = H.or.(V.and.B)

V = V.and..not.A

Here the left hand side gives the result at time t+1, as determined by the status at time t of the variables on the right hand side. Thus the crucial last equation means that the virus survives only if there is virus and no antibody present. (As an alternative to the last equation, Neumann[6] suggests V = V.and..not.(A.and.H).)

Starting from any of the $2^5 = 32$ possible configurations, we end up after some iterations in one of five fixed points: (ASHBV) = (00000), (01100), (01110), (01000), and (01001). The first three are healthy and correspond to the virgin , the vaccinated, and the immune state; if we add antigen V to any of these three fixed points, we return to the immune state (01110). The two other fixed points correspond to paralysis of the immune system, with (01001) or without (01000) antigen; addition of virus results in paralysis with antigen, leading to permanent sickness because of this failure of the immune system.

If we now put this model onto a square lattice with nearest neighbor interactions, we first have to clarify how different lattice neighbors interact with each other. High-speed multi-spin coding simulations are easiest if one first takes for each cell type the logical OR of the five lattice neighbors involved, and regards the outcome as the variable to be used on the rhs of the above equations. For example, one regards the virus present at a lattice site if at the previous time step at least one of the five lattice sites contained a virus at the previous time step; the same rules applies to the antibodies; and finally, from to V(t+1) = V and not A, the new value of V is determined.

The result of these lattice simulations, if we start with a small antigen concentration, is the immune fixed point (01110) everywhere, which is healthy but boring. The time needed to reach this fixed point increases as the square root of the logarithm of the system size, since the process has to fill the largest hole between two initial virus [5]. More interesting results were found 6 when positive thresholds were used, i.e. when just one true out of five neighbors no longer makes the corresponding variable true. Majority rules are helpul in this and other cases to get stable clusters or localized fluctuations, as shown in fig. 1. In the AIDS models to be discussed below[7], nontrivial end states can be found on lattices already

Model	A	B	D	H	K	L	M	O	S	V	refs.
Extended KUT model	1	1	0	1	0	0	0	0	1	1	(3),4,6
3-cell auto-immune	0	0	0	1	1	0	0	0	1	0	4
5-cell auto-immune	0	0	0	2	1	0	0	0	2	0	8,5,9,4
3-cell AIDS model	0	0	1	1	0	0	0	0	1	0	7
4-cell AIDS model	0	0	1	1	0	0	1	0	1	0	7
8-cell AIDS model	1	0	2	1	0	1	1	0	2	0	7
Unified model for all	1	1	1	1	1	0	0	1	1	1	10

Table: Summary of immunological models. We give the number of spins used for each kind of cell type: A: antibodies, B: bone-marrow derived lymphocytes, D: AIDS virus or other destructive agent, K: killer cells attacking own body O, L: lymphokines, M: macrophages, H: helper cells, O: self-antigen (own body attacked by killer cells), S: suppressor cells, V: antigen (e.g. virus)

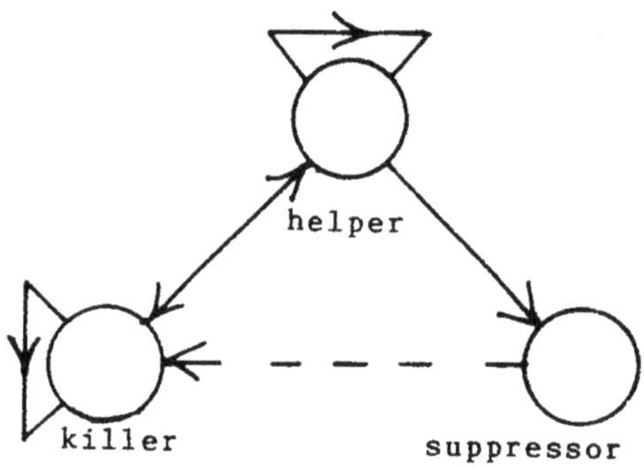

Fig.1: Schematic representation of the simple model for auto-immune disease[4].

without such tricks.

Simulations in the physics of condensed matter usually deal with only a small fraction of a real system, and often only with very short times. An immune system, on the other hand, consists of about 10^9 cells. Thus a 9600 * 9600 lattice, with five cell variables on each lattice site, is already of the size of a natural system. Then, a Cray processor makes each second a time step corresponding to 10^2 hours in real immune systems. Thus if our models would be realistic, we could simulate Nature quite nicely.

IMMUNE DISEASES

First we deal with the case where the immune system attacks the own body (auto-immune disease). The simplest case involves killer cells K which attack, for example, the own nerves, helper cells H and suppressor cells S. Between each type of cells one can have a supporting, a diminishing, or no interaction, giving 3^{n*n} possible models with n cells; thus for n=3 we get 19683 possibilities. Going through all of them, omitting unrealistic cases where helpers suppress etc, and also omitting cases where two cell types have the same behavior, and requiring exactly three fixed points healthy (KHS = 000), immune (KHS = 011), and sick (KHS = 111), Chowdhury [4] arrived at exactly one model, depicted schematically in fig.1.

H = H.or.K

$K = K + H > S$

S = H

It would be somewhat difficult to repeat such a systematic search for systems of differential equations or other more complicated methods. (Again, the RHS is valid at time t+1 if the RHS is the configuration at time t.) In the equation for K, the arithmetics on the RHS means that we treat the Boolean variables K, H, and S also as integers 0 and 1. K is set true or 1 if the sum K+H is larger than S; otherwise K is false, K=0.) On a lattice, starting with a small concentration of helper cells, the whole lattice fills up with immune lattice sites. The time needed to reach this fixed point increases as the square root of the logarithm of the system size, since the process has to fill the largest hole between two initial virus 5. When we change the threshold conditions on a lattice, we may get the clusters and oscillations shown in our fig.2.

```
7 7 7 7 7 0 0 0 0 0 0 0 0 0 0 0 0 0 0 0 0 0 0 0 0 0 0 0 0 0 0 0 0 7 7 7 7 7 7
7 7 7 7 7 0 0 0 0 0 0 0 0 0 0 0 0 0 0 0 0 0 0 7 7 7 7 7 0 0 0 0 0 7 7 7 7 7 7
7 7 7 7 7 0 0 0 0 0 0 0 0 0 0 0 0 0 0 0 0 0 0 7 7 7 7 7 0 0 0 0 0 7 7 7 7 7 7
7 7 7 7 7 0 0 0 0 0 0 0 010 910 0 0 0 0 0 0 0 7 7 7 7 7 0 0 0 0 0 7 7 7 7 7 7
7 7 7 7 7 0 0 0 0 7 7 0 0 9 8 9 0 0 0 0 0 0 0 7 7 7 7 7 0 0 0 0 0 7 7 7 7 7 7
7 7 7 7 7 0 0 0 0 7 7 0 0 8 9 8 0 0 0 0 0 0 0 7 7 7 7 7 0 0 0 0 0 7 7 7 7 7 7
7 7 7 7 7 0 0 0 0 0 0 0 0 0 8 9 0 0 0 0 0 0 0 7 7 7 7 7 0 0 0 0 0 7 7 7 7 7 7
7 7 7 7 7 0 0 0 0 0 0 0 0 0 010 0 0 0 0 0 0 0 7 7 7 7 7 0 0 0 0 0 7 7 7 7 7 7
7 7 7 7 7 0 0 0 0 0 0 0 0 0 0 0 0 0 0 0 0 0 0 7 7 7 7 7 0 0 0 0 0 7 7 7 7 7 7
7 7 7 7 7 0 0 0 7 7 7 7 0 0 0 0 0 0 0 0 0 0 0 7 7 7 7 7 0 0 0 0 0 7 7 7 7 7 7
7 7 7 7 7 0 0 0 7 7 7 7 0 0 0 0 0 0 0 0 0 0 0 7 7 7 7 7 0 0 0 0 0 0 0 0 0 0 0
0 0 0 0 0 0 0 0 7 7 7 7 0 0 0 0 0 0 0 0 0 0 0 7 7 7 7 7 0 0 0 0 0 0 0 0 0 0 0
0 0 0 0 0 0 0 0 0 0 0 0 0 0 0 0 0 0 0 0 0 0 0 7 7 7 7 7 0 0 0 0 0 0 0 0 0 0 0
0 0 0 0 0 0 0 0 0 0 0 0 0 0 0 0 0 0 0 0 0 0 0 7 7 7 7 7 0 0 0 0 0 0 0 0 0 0 0
0 0 0 0 0 0 0 0 0 0 0 0 0 0 0 0 0 0 0 0 0 0 0 7 7 7 7 7 0 0 0 0 0 0 0 0 0 0 0
0 0 0 0 0 0 0 0 0 0 7 7 7 7 0 0 0 0 0 0 0 0 0 0 0 0 0 0 0 0 0 0 0 0 0 0 0 0 0
0 0 0 0 0 0 0 0 0 0 7 7 7 7 0 0 0 0 0 0 0 0 0 0 0 0 0 0 7 7 7 0 0 0 0 0 0 0 0
0 0 0 0 0 0 0 0 0 0 7 7 7 7 0 0 0 0 0 0 0 0 0 0 0 0 0 0 7 7 7 0 0 0 0 0 0 0 0
0 0 0 0 7 7 7 7 0 0 0 0 0 0 0 0 0 0 0 0 0 0 0 0 0 0 0 0 7 7 7 0 0 0 0 0 0 0 0
0 0 0 0 7 7 7 7 0 0 0 0 0 0 0 0 0 0 0 0 0 0 0 7 7 7 0 0 7 7 7 0 0 0 0 0 0 0 0
0 0 0 0 0 0 0 0 0 0 0 0 0 0 0 0 0 0 0 0 0 0 0 7 7 7 0 0 0 0 0 0 0 0 0 0 0 0 0
0 0 0 0 0 0 0 0 0 0 0 0 0 0 0 0 0 0 0 0 0 0 0 7 7 7 0 0 0 0 0 0 0 0 0 0 0 0 0
0 0 0 0 0 0 0 0 0 0 0 0 0 0 0 0 0 0 0 0 0 0 0 0 0 0 0 7 7 7 7 7 0 0 0 0 0 0 0
0 0 0 0 0 0 0 0 0 0 0 0 0 0 0 0 0 0 0 0 0 0 0 0 0 0 0 7 7 7 7 7 0 0 0 0 0 0 0
0 0 0 0 7 7 0 0 0 0 7 7 7 7 7 0 0 0 0 0 0 0 0 0 0 0 0 7 7 7 7 7 0 0 0 0 0 0 0
0 0 0 0 7 7 0 0 0 0 7 7 7 7 7 0 0 0 0 0 0 0 7 7 7 0 0 7 7 7 7 7 0 0 0 0 0 0 0
0 0 0 0 7 7 0 0 0 0 7 7 7 7 7 0 0 0 0 0 0 0 7 7 7 0 0 0 0 0 0 0 0 0 0 0 0 0 0
7 7 0 0 7 7 0 0 0 0 7 7 7 7 7 0 0 0 0 0 0 0 7 7 7 0 0 0 0 0 0 0 0 0 0 0 0 0 0
7 7 0 0 0 0 0 0 0 0 7 7 7 7 7 0 0 0 0 0 0 0 7 7 7 0 0 0 0 0 0 0 0 0 0 0 0 0 0
0 0 0 0 0 0 0 0 0 0 7 7 7 7 7 0 0 0 0 0 0 0 0 0 0 0 0 0 0 0 0 0 0 0 0 0 0 0 0
0 0 0 0 0 0 0 0 0 0 7 7 7 7 7 0 0 0 0 0 0 0 0 0 0 0 0 0 0 0 0 0 0 0 0 0 0 0 0
0 0 0 0 0 0 0 0 0 0 7 7 7 7 7 0 0 0 0 0 0 0 0 0 0 0 0 0 0 0 0 0 0 7 7 7 7 7 7
7 7 7 7 7 0 0 0 0 0 0 0 0 0 0 0 0 0 0 0 0 0 0 0 0 0 0 0 0 0 0 0 0 7 7 7 7 7 7
7 7 7 7 7 0 0 0 0 0 0 0 0 0 0 0 0 0 0 0 0 0 0 0 0 0 0 0 0 0 0 0 0 7 7 7 7 7 7
7 7 7 7 7 0 0 0 0 0 0 0 0 0 7 7 7 0 0 0 0 0 0 0 0 0 0 0 0 0 0 0 0 7 7 7 7 7 7
7 7 7 7 7 0 0 0 0 0 0 0 0 0 7 7 7 0 0 0 0 0 0 0 0 0 0 0 0 0 0 0 0 7 7 7 7 7 7
7 7 7 7 7 0 0 0 0 0 0 0 0 0 0 0 0 0 0 0 0 0 0 0 0 0 0 0 0 0 0 0 0 7 7 7 7 7 7
7 7 7 7 7 0 0 0 0 0 0 0 0 0 0 0 0 0 0 0 0 0 0 0 0 0 0 0 0 0 0 0 0 7 7 7 7 7 7
7 7 7 7 7 0 0 0 0 0 0 0 0 0 0 0 0 0 0 0 0 0 0 0 0 0 0 0 0 0 0 0 0 7 7 7 7 7 7
```

Fig.2: Clusters of sick sites (7) in a sea of healthy sites (0) for a modified three-cell model of auto-immune disease. Numbers above 7 refer to oscillating sites. While such lattice models connect us to usual cellular automata they may be less relevant biologically if the immune response is slower than the spreading of diseases etc within the body.

More realistic are five-cell models, with two types of helpers, two types of suppressors, and the active killers. The first model, of Weisbuch and Atlan [8], had great influence on the later developments of ref.4-7, but Cohen and Atlan [9] suggested several modifications. Going again through all possible five-cell models and imposing many biologically motivated restrictions, Chowdhury [4] found one model giving only the three fixed points healthy (0000), immune, and sick:

K = K + H1 + H2 > S1 + S2

H1 = H1 + H2 > S1

H2 = K.or.H1

S1 = H1.or.H2

S2 = H1

On a square lattice, all sites become permanently sick after a short time.

The AIDS disease is due to a virus HIV and weakens the helper cell populations until the immune system is no longer working; the body may then no longer defend itself against normally non-lethal diseases like pneumonia. Pandey[7] studied several cellular automata models for the beginning stages of AIDS; we review here only the one of intermediate complexity with four cells: virus D, helpers H, suppressors S, and macrophages M. Macrophages are important since the AIDS virus can hide there; in the model, M refers only to those macrophages which have the virus inside. The dynamics of the model is:

D = (D.or.H.or.M).and..not.S

H = (H.or.M).and..not.D

S = V.and.H.and.M

M = M.and.D

It leads, depending on the initial configuration, to one of five final states for the four concentrations (DHSM): fixed point healthy (0000), fixed point sick (1001), and a cyclus of period 3: from susceptible (0001) to infected (1101) to strongly infected (1011) and back to susceptible. On a simple cubic lattice, all sites acquire after some time macrophages and suppressor, whereas the number of virus and helpers oscillate and depend strongly on the initial concentration.

One can also mix two different interactions (also for autoimmun diseases) in which case either each site selects at the beginning which rule it wants to follow (quenched disorder), or makes the choice randomly again at each different iteration (annealed disorder). The two types of disorder give qualitatively similar results. So far, the percolation threshold did no show up as a sharp phase transition in these mixture models.

Finally, one may try a unified model, giving both the normal immune response and the above immune diseases. Enlarging[10] the extended KUT model, described above for the healthy case, by two fixed parameters D and O, corresponding to the AIDS virus and the self-antigen (e.g. nerves attacked by the activated killers K of auto-immune diseases), adding for the killer cells K = (O.and.H).and..not.S to the equations, and modifying the helper equation to H = (H.or.((O.or.V).and..not.S).and..not.D, we recover the normal immune response without self-antigen and without AIDS (O = D = false). The presence of the AIDS virus, on the other hand, destroys the helper cells and makes the body vulnerable to subsequent virus attack (V = true). On a lattice, one can simulate how an initial infection (V = true) leads to immunity, which then can be destroyed by AIDS. A following infection (V = true) then no longer will be suppressed by the immune system.

1. Cohen, I.R., in: *Theories of Immune Networks*, eds. Atlan, H. and Cohen,I.R., Springer Verlag, Heidelberg 1989.

2. Atlan H., Bull.Math.Biology 51, 247 (1989).

3. Kaufman,M., Urbain,J. and Thomas,R., J.Theor.Biology 114, 527 (1985)

4. Chowdhury, D. and Stauffer, D., J.Stat.Phys, submitted (1989).

5. Dayan,I., Stauffer,D., and Havlin,S., J.Phys. A 21, 2473 (1989); Wiesner,U., J.Undergrad.Res.in Phys. 7, 15 (1989).

6. Neumann,A., Physica A, in press (1989)

7. Pandey, R.B., J.Stat.Phys. 54, 997 (1989), J.Theor.Biol., submitted (1989); J.Phys. A, submitted (1989).

8. Weisbuch,G. and Atlan,H., J.Phys. A 21, L 189 (1989)

9. Cohen,I.R. and Atlan,H., J.Autoimmunity, in press (1989)

10. Chowdhury,D., Stauffer, D., and Choudary, P.V., preprint (1989)

FINDING THE LINEAR INVARIANTS OF LATTICE GASES

D. d'Humières, Y.H. Qian, and P. Lallemand
Laboratoire de Physique Statistique de l'École Normale Supérieure,
C.N.R.S. and Université Pierre et Marie Curie,
24 rue Lhomond, 75231 Paris Cedex 05, FRANCE

ABSTRACT

Hydrodynamical phenomena can be simulated by discrete lattice gas models obeying cellular automata rules with suitable restrictions on the crystallographic symmetries of the underlying lattice. However, the derivation of the dynamical equations assumes that mass, momentum and eventually energy are the only conserved quantities during the time evolution of the automata. We shall present here an exact method to find all the conserved linear quantities of any given model, method which is equivalent to find the null space of a matrix derived from the time evolution of the automaton. Thus, the algorithmic complexity is of order N^3, where N is the number of nodes of the lattice. We shall also give a faster method when the model is invariant under the translation group of the lattice. These results will be applied to several one- and two-dimensional models. Finally, we shall note that this method gives the linear invariants of any cellular automaton.

1. INTRODUCTION.

A "lattice gas" is a special class of cellular automata modeling the synchronous motion of particles on a regular lattice. In 1985, Frisch, Hasslacher and Pomeau[1] have shown that a particular model (FHP) leads to the Navier–Stokes equations after proper limit is taken and under assumptions which will be clarified later. This finding has led to a rapid development of the subject and extensions of the original model to a variety of other physical problems[2].

The approximate macroscopic equations governing the dynamics of the system are derived from the underlying exact microscopic time evolution using the key assumption that there exist only very few **known** quantities which are conserved at the macroscopic level during the time evolution, namely the total mass and momentum for the Navier–Stokes equations.[1, 3–5] This assumption was thought to be true for the FHP models, but three new global invariants were also found for all them.[6] This class of invariants, which have no physical counterpart, plagues all the other models studied until now. While these invariants exist at the macroscopic level only for special sizes of the lattice and periodic boundary conditions, their local fluctuations are coupled to the hydrodynamical equations for any size and boundary. Even though the corresponding terms are damped for special geometry, their relaxation time for very large lattice size can become much larger than the hydrodynamical time scales and, then, their contributions have to be taken into account. These spurious invariants seem also to produce non–physical Knudsen layers at the boundaries, effects which are still under investigation.[7]

In a previous paper[8] we have tried to give more precise definitions of these quantities but we were unable to answer the questions we asked at that time. Recently, Zanetti[9] found a way to count all the invariants of any given model, using a method based on the Boltzmann approximation and the linearized collision matrix. In this paper, we shall present a more general approach which does not require any approximation and can be applied to any cellular automaton.

In section 2 we introduce a general class of reversible lattice gases and give a short description of some one–dimensional models,[10–12] the basic model introduced by Hardy, de Pazzis, and Pomeau (HPP)[13–15], and finally the FHP, and face–centered hypercubic[3] (FCHC) models which lead to hydrodynamical behaviours. Section 3 is devoted to the various possible definitions of an invariant, starting with the concept of "geometrical" invariants introduced in reference 8, we shall generalize it and define "global" linear invariants. This new definition will be used to find the invariants as the null space of a square matrix derived from the exact time evolution of the automaton. Using the translation group of the lattice, a faster method will be outlined in section 4 and applied in section 5 to the one–dimensional models and to the HPP and FHP models. As a conclusion, the application of these results to general cellular automata and to more complex invariants will be mentioned in section 6.

2. DETERMINISTIC LATTICE GAS MODELS.

A cellular automaton (CA) is a set of processing units laying on a regular lattice $\mathcal{L}$ at discrete positions $\mathbf{r}^\star \in \mathcal{L}$, each unit being characterized at discrete time steps $t^\star$ by an internal state $S(t^\star, \mathbf{r}^\star)$ coded with b bits. The time evolution of the automaton is given by a set of p vectors $\mathbf{c}_i$, such that $\mathbf{r}^\star + \mathbf{c}_i \in \mathcal{L}$, and a function $\mathcal{F}$ of $\{0,1\}^{bp} \to \{0,1\}^b$ relating the internal state at time $t^\star + 1$ and position $\mathbf{r}^\star$ to the ones at time $t^\star$ and positions $\mathbf{r}^\star - \mathbf{c}_i$ by:

$$S(t^\star + 1, \mathbf{r}^\star) = \mathcal{F}\left(S(t^\star, \mathbf{r}^\star - \mathbf{c}_1), \ldots, S(t^\star, \mathbf{r}^\star - \mathbf{c}_p)\right). \tag{1}$$

A lattice gas is a particular class of CA's such that $b = p$ and $\mathcal{F}$ is a set of b functions f_i of $\{0,1\}^b \to \{0,1\}$ such that:

$$s_i(t^\star + 1, \mathbf{r}^\star) = f_i\left(S(t^\star, \mathbf{r}^\star - \mathbf{c}_i)\right), \qquad i \in \{1, \ldots, b\}, \tag{2}$$

where s_i is the ith bit of S. This time evolution can be split in two steps:

- a "collision step" of zero duration, represented by a collision operator $\mathcal{C}$: $s_i(\mathbf{r}^\star) \mapsto f_i\left(S(\mathbf{r}^\star)\right)$,
- followed by a "propagation step" of duration equal to one time unit, represented by a streaming operator $\mathcal{S}$: $s_i(\mathbf{r}^\star) \mapsto s_i(\mathbf{r}^\star - \mathbf{c}_i)$.

With this representation, the dynamics of the lattice gas can be viewed as the dynamics of a gas of Boolean particles moving at discrete time steps on a regular lattice with discrete velocities $\mathbf{c}_i$. In order to recover some physical behaviour for these particles, the collision operator is chosen such that:

$$\sum_i a_{ij} f_i(S) = \sum_i a_{ij} s_i, \qquad \forall S \in \{0,1\}^b, \tag{3}$$

for q b–vectors $\{a_{ij}\}$, $i \in \{1, \ldots, b\}$ and $j \in \{1, \ldots, q\}$, which will be called microscopic invariants in the sequel. For the lattice gases used in references 1, 3–5, and 13–15, $q = D+1$, $\{a_{i1}\} = \{m_i\}$, and $\{a_{ij}\} = \{m_i c_{i\alpha}\}$, $j = \alpha + 1$, $\alpha \in \{1, \ldots, D\}$, where D is the dimensionality of the space, m_i the mass of particle i, and $c_{i\alpha}$ the component of $\mathbf{c}_i$ in the direction α. In this case, the conservation relations express the conservation of mass and momentum during the collision, as it occurs in the real world, while the conservations of mass and energy are equivalent.

These microscopic conservation laws are destroyed by the propagation steps and lead, in the ideal case, to only q macroscopic conserved quantities:

$$I_j = \sum_{i\ \mathbf{r}^\star} a_{ij} s_i(t^\star, \mathbf{r}^\star), \qquad \forall j \in \{1, \ldots, q\} \text{ and } \forall t^\star. \tag{4}$$

Assuming "molecular chaos" and "semi–detailed balance" it can be shown that the system evolves towards an equilibrium state which is completely determined by these conserved quantities. Glueing together local equilibria and using a Chapman–Enskog expansion,[16] it is then possible to derive macroscopic equations describing the dynamical behaviour of the system.

In the Chapman–Enskog expansion, symmetrical fourth–order tensors appear which must be isotropic in order to obtain the Navier–Stokes equation. This requirement is fulfilled if the set of $\mathbf{c}_i$'s and the collision operator have the same symmetry group than the lattice $\mathcal{L}$ and the lattice is either the triangular lattice in two dimensions[1] or the face–centered–hypercubic (FCHC) lattice in four dimensions[3]. The invariance of the collision operator under the symmetry group $\mathcal{G}$ of the lattice can be expressed by:

$$A\left(g(S) \rightarrow g(S')\right) = A(S \rightarrow S'), \qquad \forall g \in \mathcal{G} \tag{5}$$

where $A(S \rightarrow S')$ is the probability to have the output state S' from the input state S and verifies $\sum_{S'} A(S \rightarrow S') = 1$ (normalization) and $\sum_S A(S \rightarrow S') = 1$ (semi–detailed balance).

In what follows, it will be useful to consider reversible lattice gases which imply a deterministic collision operator and semi–detailed balance. If $\mathcal{G}$ has b_g elements, $A(S \rightarrow S')$ can always be chosen equal to k/b_g, where k is a factor of b_g. We shall assume that for any model it is possible to define an extended internal state $S_E = S \cup S_F$, where $S_F \in \{1, \ldots, b_g\}$ is varying from nodes to nodes but independent of time ($S'_F = S_F$) and such that $A(S_E \rightarrow S'_E) = 1$, if $A(S \rightarrow S') \neq 0$, and $\sum_{S_F} A(S_E \rightarrow S'_E) = b_g A(S \rightarrow S')$ for all S_E and S'_E. These conditions ensure a reversible dynamics and a reasonable approximation of the probabilities, if b_g is smaller than the number of nodes used for local averaging.

2.1 One–Dimensional Models.

The simplest lattice gas is a three–bit one–dimensional model, having two particles with mass 1 and velocities ± 1 and a rest particle with mass 2. The only collisions, which conserve mass and momentum, are the "head–on" collision of two moving particles giving the rest particle, and the splitting of the rest particle into the moving particles with opposite velocities.[10]

There exists a four–bit model with conservation of mass, momentum and energy. Two "hot" particles have mass 1 and velocities ± 2, and two "cold" ones have mass 2 and velocities ± 1. The only possible collisions are the collisions of one hot and one cold particles with opposite velocities.[11]

We shall show later that these models have spurious invariants while the following model has only mass and momentum conservations. It consists of 5 particles with mass 1 and velocities in the set $\{-2,-1,0,1,2\}$. In this case, six pairs of states can be exchanged with mass and momentum conservations: $\{1,0,0,0,1\}$ and $\{0,1,0,1,0\}$, $\{1,0,0,1,0\}$ and $\{0,1,1,0,0\}$, $\{0,1,0,0,1\}$ and $\{0,0,1,1,0\}$ and the corresponding pairs where the 0 and 1 are exchanged.[12]

2.2 The HPP Model.

The simplest 2–D model, introduced by Hardy, de Pazzis and Pomeau[13–15], consists of particles moving on a square lattice with one of the following velocities $\{1,0\}$, $\{0,1\}$, $\{-1,0\}$, and $\{0,-1\}$. The only states leading to an effective collision process conserving mass and momentum, are the two "head–on" situations where exactly two particles are present with opposite velocities, these states being exchanged during the collision step. This model gives dynamical equations which are not invariant under arbitrary rotations as soon as second order terms in the Chapman-Enskog expansion are involved; thus the HPP model is not suitable to observe hydrodynamical behaviours.

2.3 The FHP Models.

The FHP model consists of particles moving on a triangular lattice with velocities $\{\cos(2k\pi/3), \sin(2k\pi/3)\}$, $k \in \{1,\ldots,6\}$ (and possibly with a seventh particle at rest). For this model, there exist several sets of states which can be exchanged during a collision process with mass and momentum conservation. Thus we cannot define a unique FHP model as it was done for the HPP one; however, some care must be taken in the choice of the collisions to avoid extra invariants at the microscopic level, as it was done in the original FHP model (FHP I) with only head–on and symmetric three–body collisions. Moreover, there exist situations for which an input state can be exchanged with several other ones, such that the head–on collisions. In this paper, we shall assume that an extra bit is added to each state which allows one of the two possible output states to be chosen (the input configuration is rotated clockwise if this bit is zero and counter–clockwise otherwise, $S_F \in \{0,1\}$). This extra bit is set at random with probability 1/2 and the collision rules will be chosen such that the dynamics of the lattice gas is reversible under time reversal (the extra–bit is complemented when time is reversed).

2.4 The Face–Centered–Hypercubic 4–D And The Pseudo–4–D Models.

The FCHC model consists of 24 particles moving on a hypercubic lattice in four dimensions with velocities $\{\pm1,\pm1,0,0\}$, $\{\pm1,0,\pm1,0\}$, $\{\pm1,0,0,\pm1\}$, $\{0,\pm1,\pm1,0\}$, $\{0,\pm1,0,\pm1\}$, $\{0,0,\pm1,\pm1\}$. This model can be projected down to three dimensions to recover the standard 3–D Navier–Stokes equation. For this model the number of states which can be exchanged during the collision process is enormous[17–19] and, at this time, the minimum number of output states required to obtain collision rules with the same symmetries as the lattice is not known (this number is smaller or equal to 1152), neither the possibility to build a reversible model invariant under the symmetry group of the lattice (when needed we shall assume the existence of such a model).

3. INVARIANTS IN DETERMINISTIC LATTICE GASES.

3.1 Geometrical Invariants.

For a lattice of size N_α in direction α, for a total of $N = \prod_\alpha N_\alpha$ nodes, and periodically wrapped around, the previous analysis relies on the competition between the collision steps and the propagation steps to reduce the number of conserved quantities from qN after the collisions, to exactly q when both steps are completed. Indeed, such a drastic reduction in the number of conserved quantities requires some kind of "miracle" which cannot occur for every model. For example, it was known since the early works of HPP that the component along x (resp. y) of the momentum summed on any line parallel to the x (resp. y) axis is conserved during the time evolution of the HPP model. It was also believed that this miracle occurred for the FHP and FCHC models, but this hope was shown to be false,[6] and it does not exist yet a rigorous proof of the existence of lattice gas models with only $D+1$ invariants leading to the Navier–Stokes equations without coupling to spurious dynamical equations.

A first definition of invariants comes from the theory of reversible cellular automata of finite size. In this case the number of accessible states is finite and, because each state has only one ancestor and one successor, for any initial configuration the time evolution is a closed trajectory in phase space. These trajectories can be labelled by integers which do not change during the time evolution of the CA and can be used to define the invariants of reversible lattice gases.

However, from a physical point of view, this definition is not very useful because for most of the systems of interest there exist unstable configurations:

situations with a peculiar dynamics but which are destroyed by the perturbation of a single bit of the lattice internal state. For this reason, we shall only consider in the sequel, invariants somehow independent of the initial configurations.

In reference 8, we have defined two kinds of such invariants, that we called "geometrical" invariants:

- a ***static invariant*** is the union of a b–vector $\mathbf{a}$, a sublattice $\mathcal{I}$ of $\mathcal{L}$ and an integer $T^\star$, independent of the size of the lattice $\mathcal{L}$, such that:

$$\sum_{i\ \mathbf{r}^\star \in \mathcal{I}} a_i s_i(t^\star + T^\star, \mathbf{r}^\star) = \sum_{i\ \mathbf{r}^\star \in \mathcal{I}} a_i s_i(t^\star, \mathbf{r}^\star) , \qquad \forall t^\star \text{and } s_i(0, \mathbf{r}^\star) ; \quad (6)$$

- a ***moving invariant*** is the union of a b–vector $\mathbf{a}$, a sublattice $\mathcal{I}$ of $\mathcal{L}$ and a translation operator $\mathcal{T}$ such that:[1]

$$\mathcal{T}\mathcal{I} \neq \mathcal{I} , \ \mathcal{T}\mathcal{I} \subset \mathcal{L} ,$$
$$\text{and} \quad \sum_{i\ \mathbf{r}^\star \in \mathcal{T}\mathcal{I}} a_i s_i(t^\star + 1, \mathbf{r}^\star) = \sum_{i\ \mathbf{r}^\star \in \mathcal{I}} a_i s_i(t^\star, \mathbf{r}^\star) , \quad \forall t^\star \text{and } s_i(0, \mathbf{r}^\star) . \quad (7)$$

The requirement that $T^\star$ is independent of the size of the lattice is introduced in the definition of the static invariant to prevent the confusion between a moving invariant and a static one with $T^\star$ of the order of the minimum value of the N_α.

3.2 Global Linear Invariants.

Let us call *global linear invariant* with a time period $T^\star$, an N_b–vector $\mathbf{a}_T = \{a_{Ti}(\mathbf{r}^\star)\}$ such that:

$$\sum_{i\ \mathbf{r}^\star} a_{Ti}(\mathbf{r}^\star) s_i(kT^\star, \mathbf{r}^\star) = \sum_{i\ \mathbf{r}^\star} a_{Ti}(\mathbf{r}^\star) s_i(0, \mathbf{r}^\star) , \qquad \forall k \text{ and } s_i(0, \mathbf{r}^\star) , \qquad (8)$$

where $N_b = bN$ is the dimension of the phase space. The relations between the global invariants with static and moving ones will be given in the next section. Equation (8) can be written:

$$\sum_{i\ \mathbf{r}^\star} a_{Ti}(\mathbf{r}^\star) \left(s_i(kT^\star, \mathbf{r}^\star) - s_i(0, \mathbf{r}^\star) \right) = 0 , \qquad \forall k \text{ and } s_i(0, \mathbf{r}^\star) . \qquad (9)$$

---

[1] The translation and isometry operators are assumed to be defined in a way compatible with periodic boundary conditions: $\mathcal{T}\mathcal{L} = \mathcal{L}$, $\forall \mathcal{T} \in \mathcal{T}^\star$, and $g\mathcal{L} = \mathcal{L}$, $\forall g \in \mathcal{G}$, where $\mathcal{G}$ and $\mathcal{T}^\star$ are respectively the symmetry and discrete translation groups.

For a given $T^\star$ and $k \in \{1,\ldots,N_b\}$, one gets a set of N_b equations with N_b unknowns which determines the values of the $a_i(\mathbf{r}^\star)$ for a given set of $s_i(0,\mathbf{r}^\star)$. Let us define the corresponding $N_b \times N_b$ matrix by:

$$A_{lk} = s_i(kT^\star, \mathbf{r}^\star) - s_i(0, \mathbf{r}^\star) \; , \tag{10}$$

where l is a composite index running through the phase space: $l = \mathbf{r}^\star + (i-1)N$, for $\mathbf{r}^\star \in \{1,\ldots,N\}$ and $i \in \{1,\ldots,b\}$. Then, either this matrix is non singular, $a_{T_i}(\mathbf{r}^\star) = 0$, and there is no invariant, or the matrix is singular and its null space gives the set of global linear invariants with a period $T^\star$ in time. Since the elements of the matrix A_{lk} are integers, the values of the components of a global invariants can be restricted to integer values without loss of generality.

The recipe to find all the global invariants for any given $T^\star$ is now straightforward:

- pick at random an initial condition $s_i(0,\mathbf{r}^\star)$,
- iterate the lattice gas $N_b T^\star$ time steps, storing the configuration every $T^\star$ steps (the choice of $T^\star$ will be explained later),
- compute the matrix $\{A_{lk}\}$,
- find the null space of this matrix.

As shown in the appendix, the computational complexity of this algorithm is cubic in the size of the lattice and can be used on standard mid–size computers for a maximum number of nodes between 100 and 1000 depending upon the value of b and the available CPU time.

This algorithm is exact and does not depend on any hypothesis commonly used in this field, such as reversibility, detailed or semi-detailed balance, Boltzmann approximation, linearization of the collision operator, ... It gives all the global invariants for a given $T^\star$ and a given initial condition. In order to obtain a set independent of initial conditions, several runs must be made with new initial conditions. Since the cost of finding the null space of large matrices is high, better solutions are obtained by picking the invariants of a given initial condition and checking if they remain invariants for new initial ones, or by replacing in definition (10) the s_i by their average value for several initial conditions. We used the latter method for our simulations by packing 64 systems in one computer word and using bitwise logical operations to compute the collision rules.[20]

As described, the method should be used again and again for each new value of $T^\star$, starting with $T^\star = 1$. However, since the invariant set for $T^\star = T_1^\star T_2^\star$

contains the union of those obtained for $T_1^\star$ and $T_2^\star$ and the iteration cost varies only as the number of nodes times $T^\star$, it is cheaper to pick a value of $T^\star$ as the product of the smallest common multiple of the N_α by some power of small prime numbers to get at once all the static and moving invariants for a large number of periods. The likelihood to find an invariant, having a period prime with the size of the lattice and the first prime numbers, seems very low from the simulations we have done, but this result remains to be checked more carefully. The general solution is obtained if equation (8) is written as a generalized eigensystem problem:

$$Ax = \lambda Bx, \tag{11}$$

with $A_{lt^\star} = s_i(t^\star, \mathbf{r}^\star)$, $B_{lt^\star} = s_i(t^\star - 1, \mathbf{r}^\star)$, and $x_l = a_i(\mathbf{r}^\star)$, for $t^\star \in \{1, \ldots, N_b\}$ and l being the same composite index than in definition (10). Then, the eigenvectors x of equation (11), such that $\lambda^{T^\star} = 1$ are the linear invariants with period $T^\star$. Thus, this generalized eigensystem gives all the linear invariants as the eigenvectors associated to the complex eigenvalues with modulus equal to 1, but this method is more difficult to use with very large matrices than the null space problem.[2]

In most simulations, there are special points where the microscopic rules are changed to implement obstacles, body forces, input and output conditions, ... These points destroy most of the invariants of the plain periodic model, but some new states can be added to the system at these special points for some book–keeping of the non–conserved microscopic quantities. In this case, the above results can be applied to the extended phase space. However, the limited number of nodes which can be studied with this general method makes this remark of little interest for practical applications requiring typically several million nodes.

Another problem encountered with this algorithm is the difficulty to replace the found invariants by linear combinations of them which are easier to interpret in term of physical quantities. Also, if there exist two invariants $\mathbf{a}_{T_0}$ and $\mathbf{a}'_{T_0}$ for a given $T_0^\star$, with corresponding initial values S and S' of the right hand side of equation (8), then $S'\mathbf{a}_{T_0} - S\mathbf{a}'_{T_0}$ is also an invariant with an initial value equal to zero, which appears for any time period $T^\star$. Such combinations can only be detected by comparison of different runs with different initial conditions, since $S'\mathbf{a}_{T_0} - S\mathbf{a}'_{T_0}$ depends upon the initial conditions.

[2] There is a problem of numerical accuracy to select the invariants and it will probably be needed to select the eigenvalues with a modulus close to one and check if the corresponding eigenvectors are invariants.

4. TRANSLATION INVARIANCE.

In the previous section, we did not assume any geometrical properties for the model. In this section, we shall assume that there are no special points like obstacles, and then consider the lattice invariant under its symmetry group $\mathcal{G}$ and its discrete translation group $\mathcal{T}^*$. The latter invariance can be used to vastly improve the computational efficiency of the method of section 3. The full proof of the following derivation is quite long and is left for a forthcoming paper. We shall only sketch here the solution for one and two dimensional lattices with N_α equal to some power of two. In what follows, we shall assume that we are searching all the invariants with a given period in time T^*.

4.1 General Properties.

Let us define the translation operator $\mathcal{T}$, acting on an N_b–vector $\mathbf{a} = \{a_i(\mathbf{r}^*)\}$ of phase space and associated to a translation $\mathcal{T}$ in lattice space, by:

$$\mathcal{T} a_i(\mathbf{r}^*) = a_i(\mathcal{T}\mathbf{r}^*) \, . \tag{12}$$

Since the global invariants do not depend upon the initial condition, they are invariant for any geometrical transformation leaving the lattice unchanged, and, if $\mathbf{a}_T$ is a global invariant, then $\mathcal{T}\mathbf{a}_T$ is a different one for any translation $\mathcal{T}$ such that $\mathcal{T}\mathbf{a}_T \neq \mathbf{a}_T$.

Let $\mathcal{T}_\alpha$ be the translation moving the lattice one node in the direction α, and $\mathcal{T}_\alpha$ the associated translation in phase space, a global invariant has a spatial period l_α in the direction α iff:

$$\begin{aligned} \mathcal{T}_\alpha^l \mathbf{a}_T \neq \mathbf{a}_T, & \quad \forall l \text{ such that } 0 < l < l_\alpha \, , \\ \text{and} \quad \mathcal{T}_\alpha^{l_\alpha} \mathbf{a}_T = \mathbf{a}_T \, . & \end{aligned} \tag{13}$$

For periodic boundary conditions, $\mathcal{T}_\alpha^{N_\alpha}$ is equal to the identity operator and it is easy to check that the period l_α must divide N_α.

For any global linear invariant $\mathbf{a}_T$ with a period l_α in the direction α, then:

1) the l_α vectors $\mathcal{T}_\alpha^l \mathbf{a}_T$ with $0 \leq l < l_\alpha$ build up a set of global linear invariants,
2) $\sum_{l=0}^{l_\alpha - 1} \mathcal{T}_\alpha^l \mathbf{a}_T$ is also an invariant, which is uniform in the direction α, if it is non zero,

3) and, if l_α is not a prime factor of N_α: $l_\alpha = l_{\alpha 1} l_{\alpha 2}$,

$$\sum_{l=0}^{l_{\alpha 2}-1} T_\alpha^{l \cdot l_{\alpha 1}} \mathbf{a}_T \qquad \text{and} \qquad \sum_{l=0}^{l_{\alpha 1}-1} T_\alpha^{l \cdot l_{\alpha 2}} \mathbf{a}_T$$

are also invariants, with period $l_{\alpha 1}$ and $l_{\alpha 2}$ respectively, if they are not null vectors.

The above results can now be used to derive a hierarchical algorithm to find the global linear invariants for a problem invariant under the translation group of the lattice.

- Find the homogeneous invariants which are indeed the microscopic ones.
- For all the prime factors l_α of N_α, find the invariants with spatial period l_α in the direction α which are orthogonal to the homogeneous ones.
- For all the prime numbers $l_{\alpha 1}$ and $l_{\alpha 2}$ such that $l_{\alpha 1} l_{\alpha 2}$ is a factor of N_α, find the invariants with spatial period $l_{\alpha 1} l_{\alpha 2}$ in the direction α, which are orthogonal to all the previous ones.
- Repeat this process until the factors of N_α are exhausted.
- When $D > 1$, repeat the whole process for all the combinations of the different directions.

At this stage, we can summarize the above process as a partitionning of the vectorial phase space in subspaces which are invariant under the translation group, each vector of one subspace being orthogonal to all the vectors of the other subspaces. Indeed, this method is yet far from an actual implementation which will be made explicit in the next two subsections for $N_\alpha = 2^{p_\alpha}$.

4.2 One–Dimensional models.

Let us consider a one–dimensional lattice with length $N = 2^p$. In this case, the above algorithm looks for invariants with spatial periods 1, 2, 4,..., 2^p, which are orthogonal. Let us define the following set of N–vectors $\mathbf{v}_K$ with components given by:

$$v_0(x^\star) = 1 \,,$$

$$v_{2^k+m}(x^\star) = \sum_{j=0}^{N/2^{k+1}} \left(\delta_{x^\star - m, j2^{k+1}} - \delta_{x^\star - m,(2j+1)2^k}\right) \,, \qquad (14)$$

$$x^\star \in \{0, \ldots, N\} \;\; k \in \{0, \ldots, p-1\}, \;\; m \in \{0, \ldots, 2^k - 1\} \,.$$

It is quite easy to check that:

- there are N such vectors,
- each of them is orthogonal to the others,
- for a given k, the spatial period is 2^{k+1},
- for a given k, the vector indexed by $m+1$ is equal to the one indexed by m translated one position in the x direction.

These vectors build an orthogonal basis of the space R^N with all the properties required in the previous subsection for one dimension and a size which is a power of two. This basis is a modification of the well known Hadamard one[21].

Let us now define the projection $\hat{\mathbf{a}}(K)$ of an N_b–vector $\mathbf{a}$ on the vector $\mathbf{v}_K$ as a b–vector with components given by:

$$\hat{a}_i(K) = \sum_{x^\star} v_K(x^\star) a_i(x^\star) . \tag{15}$$

Let $\mathbf{a}_K^{(p)}$ be the N_b–vector associated to a projection $\hat{\mathbf{a}}(K)$ by:

$$a_{Ki}^{(p)}(x^\star) = v_K(x^\star)\hat{a}_i(K) . \tag{16}$$

Using properties 1 to 3 of subsection 4.1 and the fact that the components of an invariant $\mathbf{a}_T$ are integers, it is possible to show that the vector associated to a non–zero projection of a global linear invariant is another one (possibly the same). Thus the equation (9) can be written:

$$\sum_i \hat{a}_{Ti}(K) \sum_{x^\star} v_K(x^\star)(s_i(jT^\star, x^\star) - s_i(0, x^\star)) = 0, \ \forall j \text{ and } s_i(0, x^\star) . \tag{17}$$

For a given $T^\star$ and $j \in \{1, \ldots, b\}$, there are now N sets of b equations with b unknowns, and the union of the null spaces of the N matrices $\{A_{ij}(K)\}$, where:

$$A_{ij}(K) = \sum_{x^\star} v_K(x^\star) \left(s_i(jT^\star, x^\star) - s_i(0, x^\star)\right) , \tag{18}$$

is the set of the projections of the invariants on the vectors $\mathbf{v}_K$, and the associated vectors are the global linear invariants with time period $T^\star$.

The projection needed to compute the matrices $\{A_{ij}(K)\}$ is a modification of the Walsh–Hadamard transform[22] and can be computed in a time of order N using an algorithm derived from the standard Fast Fourier Transform (FFT, see reference 23, pp. 381–429). With this method, the computational cost is linear for the static invariants, and quadratic for the moving ones.

4.3 Two–Dimensional Models.

In the 2–D case, it is not enough to use the transformation given in the previous subsection for the two directions x and y because their combination detects only the translation invariance along the horizontal and vertical directions and not the other ones, for instance, the diagonals of the lattice. When the dimensions are equal to powers of two, the algorithm is the following:

1) do the transformation in the direction x for all the vertical lines;
2) do the transformation in the direction y for all the horizontal lines.

After these two transformations, the N matrices $\{A_{ij}\}$ are given by:

$$A_{ij}(K_x, K_y) = \sum_{\mathbf{r}^\star = \{x^\star, y^\star\}} v_{K_x}(x^\star) v_{K_y}(y^\star) (s_i(jT^\star, \mathbf{r}^\star) - s_i(0, \mathbf{r}^\star)) \ . \tag{19}$$

3) For all the blocks corresponding to the invariants with period $2l_x$ and $2l_y$ both greater than 2, for $l' \in \{0, \ldots, l-1\}$, where $l = \min(l_x, l_y)$, take the sums:

$$A'_{ij}(l_x + l', l_y + l') = \sum_{m=0}^{l-1} (-1)^{m'} A_{ij}(l_x + m_x, l_y + m_y) \ , \tag{20}$$

where m', m_x, and m_y are defined by the following relations:

$$\begin{aligned} &m_x = m, \quad m_y + m' l_y = m(2l' + 1)\frac{l_y}{l_x}, \ (0 \le m_y < l_y), \quad \text{if } l_x \le l_y \ , \\ \text{or} \quad &m_x + m' l_x = m(2l' + 1)\frac{l_x}{l_y}, \ (0 \le m_x < l_x), \quad m_y = m, \quad \text{if } l_x > l_y \ . \end{aligned} \tag{21}$$

This third step is needed to take into account the invariance along directions which are neither horizontal nor vertical but connect nodes of the lattice (for instance, diagonal directions when $l_x = l_y$ and $l' = 0$ or $l' = l - 1$). It remains to be generalized for dimensions greater than 2. The union of the null spaces of the unchanged matrices $\{A_{ij}\}$, the matrices $\{A'_{ij}\}$, and those obtained by translations of equations (20) and (21) along the direction y if $l_x \le l_y$ (resp. x), gives the set of global linear invariants.

The total computational cost of this second method to find the invariants is of order N_x^{D+1} if we are looking for moving invariants, to be compared to N_x^{3D} for the first method, corresponding to a saving of the order of N_x^{2D-1} (1000, for $N_x = 10$ and $D = 2$).

4.4 Back to the Geometrical Invariants.

Writing $a_i(\mathbf{r}^\star) = a_i$ if $\mathbf{r}^\star \in \mathcal{I}$ and $a_i(\mathbf{r}^\star) = 0$ otherwise, it is obvious that any geometrical invariant is also a global linear one, and the set of geometrical invariants a subset of the global linear ones. So far, we did not find any example of a global linear invariant which was not a geometrical one, but we do not have any proof that this result is general. We have also found that the projections of the linear invariants on the modified Hadamard basis are always linear combinations of the microscopic invariants and that these projections are the same for periods 2, 4,..., $2^{l_{max}}$ in the same direction (x, y, or diagonal ones). Using the inverse modified Walsh–Hadamard transform, it is quite easy to check that the last two properties imply the equivalence between geometrical and global linear invariance. We conjecture that equation (2) (maybe (1)) implies the equivalence between the two definitions of the linear invariance.

5. NUMERICAL RESULTS.

5.1 One–Dimensional Models.

Both algorithms of sections 3 and 4 were used for the three one–dimensional models of subsection 2.1. They have given the physical invariants: mass and momentum for all the models and energy for the second one, but also staggered invariants for the first two models and no unphysical invariants for the third one.

For the first model, there is only one spurious invariant, staggered both in space and time, and related to the difference between the momentum j on the odd and even nodes of the lattice:

$$\sum_{\mathbf{r}^\star\in\mathcal{L}_o} j(t^\star+1,\mathbf{r}^\star) - \sum_{\mathbf{r}^\star\in\mathcal{L}_e} j(t^\star+1,\mathbf{r}^\star) = -\sum_{\mathbf{r}^\star\in\mathcal{L}_o} j(t^\star,\mathbf{r}^\star) + \sum_{\mathbf{r}^\star\in\mathcal{L}_e} j(t^\star,\mathbf{r}^\star)\,, \quad (22)$$

where $\mathcal{L}_o$ and $\mathcal{L}_e$ are the odd and even sublattices respectively.

For the second model, there is one spurious invariant, staggered in space, related to the difference between the number n_h of hot particles on the odd and even nodes:

$$\sum_{\mathbf{r}^\star\in\mathcal{L}_o} n_h(t^\star+1,\mathbf{r}^\star) - \sum_{\mathbf{r}^\star\in\mathcal{L}_e} n_h(t^\star+1,\mathbf{r}^\star) = \sum_{\mathbf{r}^\star\in\mathcal{L}_o} n_h(t^\star,\mathbf{r}^\star) - \sum_{\mathbf{r}^\star\in\mathcal{L}_e} n_h(t^\star,\mathbf{r}^\star)\,, \quad (23)$$

and three other invariants, staggered in space and time, one related to the difference between the number n_c of cold particles on the odd and even nodes:

$$\sum_{\mathbf{r}^\star\in\mathcal{L}_o} n_c(t^\star+1,\mathbf{r}^\star) - \sum_{\mathbf{r}^\star\in\mathcal{L}_e} n_c(t^\star+1,\mathbf{r}^\star) = -\sum_{\mathbf{r}^\star\in\mathcal{L}_o} n_c(t^\star,\mathbf{r}^\star) + \sum_{\mathbf{r}^\star\in\mathcal{L}_e} n_c(t^\star,\mathbf{r}^\star)\,, \quad (24)$$

and two related to the difference between the number of hot particles on sublattices of $\mathcal{L}_o$ and $\mathcal{L}_e$:

$$\sum_{\mathbf{r}^\star \in \mathcal{L}_{oo}} n_h(t^\star+1,\mathbf{r}^\star) - \sum_{\mathbf{r}^\star \in \mathcal{L}_{oe}} n_h(t^\star+1,\mathbf{r}^\star) = - \sum_{\mathbf{r}^\star \in \mathcal{L}_{oo}} n_h(t^\star,\mathbf{r}^\star) + \sum_{\mathbf{r}^\star \in \mathcal{L}_{oe}} n_h(t^\star,\mathbf{r}^\star) \,, \quad (25)$$

$$\sum_{\mathbf{r}^\star \in \mathcal{L}_{eo}} n_h(t^\star+1,\mathbf{r}^\star) - \sum_{\mathbf{r}^\star \in \mathcal{L}_{ee}} n_h(t^\star+1,\mathbf{r}^\star) = - \sum_{\mathbf{r}^\star \in \mathcal{L}_{eo}} n_h(t^\star,\mathbf{r}^\star) + \sum_{\mathbf{r}^\star \in \mathcal{L}_{ee}} n_h(t^\star,\mathbf{r}^\star) \,, \quad (26)$$

where $\mathcal{L}_{ee}$, $\mathcal{L}_{oe}$, $\mathcal{L}_{eo}$, and $\mathcal{L}_{oo}$ are the set of nodes with $\mathbf{r}^\star \equiv 0,\ 1,\ 2,$ and 3 (mod 4) respectively.

5.2 The HPP Model.

It is very tedious to use the algorithm given in subsection 3.2 for the HPP model when the N_α are larger than 2, because it has a large number of linear invariants and their geometrical interpretation is very difficult. For instance, if $N_x = N_y = 4$, the dimension of the invariants is 64 and there are 26 of them. Thus, we used most of the time the second method which leads to very simple vectors with four components and easy geometrical interpretation.

Using the 2–D algorithm, we have found all the invariants described in reference 8 but, at the same time, the linear relations between them. Let us count the number of invariants when $N_x = N_y = 2^p$.

- There is a checker–board invariance with two independent subsystems made of nodes with sum of coordinates odd or even.
- For each subsystem, the momentum in the direction x (resp. y) is conserved on each horizontal (resp. vertical) line.
- The sum of the number of particles moving in directions $\mathbf{c}_i$ and $\mathbf{c}_{i+1}$ along any line with direction $\mathbf{c}_i - \mathbf{c}_{i+1}$ is a moving invariant in direction $\mathbf{c}_i + \mathbf{c}_{i+1}$.
- Finally, the total number of particles is conserved for each subsystem.

Thus, there are $8N_x + 2$ invariants ($4N_x$line, $4N_x$ moving, and 2 mass invariants), but for each subsystem there are four obvious linear relations: the sum of all the moving invariants in opposite directions is equal to the total number of particles, while their difference is equal to the sum of the total momentum in the directions x and y for one pair and their difference for the other one, the number of independent invariants being then smaller or equal to $8N_x - 6$. The results of the simulations, for a wide variety of size (from 4×4 to 64×64), show that the number of independent invariants is exactly equal to $8N_x - 6$ and, thus, that all of them were already known, with no other linear relations between them than the eight ones given above.

5.3 The FHP Models.

For the FHP model, both methods give six linear invariants, the three physical ones: mass and momentum in directions x and y, and the three staggered invariants found by McNamara and Zanetti.[6] These spurious invariants correspond to the difference between the sums of the local momentum in direction $\mathbf{c}_i + \mathbf{c}_{i+1}$ on the odd and even lines in direction $\mathbf{c}_i - \mathbf{c}_{i+1}$.

5.4 Some Modified HPP and FHP Models.

We have also used both algorithms to test a very simple attempt to remove the unphysical invariants:

- for odd time steps, move particle i in direction $\mathbf{c}_i$ for the HPP model (in direction $\mathbf{c}_{i+1}$ for the FHP one)
- for even time steps, move particle i in direction $\mathbf{c}_{i+1}$ for the HPP model (resp. $\mathbf{c}_{i-1}$). This method corresponds to move particle i with an average velocity $(\mathbf{c}_i + \mathbf{c}_{i+1})/2$ (resp. $\mathbf{c}_i/2$).

For the modified FHP model, we have found only the physical invariants. For the modified HPP, we have found ten invariants: the checker–board invariance is not destroyed and for each subsystem there are the three physical invariants plus two staggered invariants corresponding to the difference between the total momentum on the odd and even horizontal lines for the directions x and y (with the checker–board invariance, the horizontal and vertical lines play a symmetrical role). In fact, each subsystem is a TM gas,[24] having only five invariants. This drastic change in the number of invariants between the HPP and TM gases may explain the qualitative differences of the velocity correlation functions found in reference 24.

We have also tested another modification of the HPP model for which the time evolution has a period four: the particle i does not move during time step i and move with velocities $\mathbf{c}_{i+1}$, $\mathbf{c}_i$, and $\mathbf{c}_{i-1}$ during the three other time steps. With this scheme we have found only the physical invariants.

These results show that appropriate modifications of the propagation rules destroy the spurious linear invariants. However, we have no proof at this time that this method does not produce non–linear spurious invariants we are unable to detect with the described algorithms. If such non–linear invariants do exist, their coupling with the hydrodynamical equations will also have to be investigated.

6. CONCLUSIONS.

Without a general algorithm to find all the linear invariants of a given model, it was rather useless to try to remove the unphysical ones of the standard models, for fear of replacing a known problem with known solutions by an unknown one. While some work remains to be done to achieve more usable algorithms to find the invariants, it is now the right time to look for models having only the desired number of invariants. One possibility may come from the study of models with modified propagation rules, as indicated in subsection 4.3. Another alternative seems to use a larger velocity set which removes the spurious invariant, as it is done for the third one–dimensional model studied here (see also reference 9). A last solution may come from the random coupling between different "universe" run in parallel.

An other subject of interest is the use of the present algorithm for general cellular automata and the search for more complicated invariants. For instance, the well known Q2R automaton[25] has an invariant, corresponding to the energy of a system of magnetic spins, which is the sum of terms $s_i s_j$.[26] This invariant, obviously non–linear, will not be found as such by the algorithm described in this paper and more work is needed to find it.

Acknowledgment.

We thank J.R. Billuart, S. Zaleski and G. Zanetti for helpful discussions. Part of this work was supported by an EEC contract number ST2J–0190.

APPENDIX: AN ALGORITHM TO FIND THE NULL SPACE.

A very simple way to solve the classical problem of finding the null space of a given square matrix can be derived from the standard method to get the inverse of a square matrix (see reference 23, pp. 19–31):

1) construct a $N_b \times N_b$ identity matrix B,
2) set `J0=1`,
3) for `I=1 to` N_b`-1` do:
    a) find the maximum value of `A(I,J)` and its position `JMAX` for `J=J0,...,`N_b,
    b) if this value is less than a given small number
    then

the matrix is considered singular and the Ith row of B gives the components of a global invariant,[3]

else

α) exchange rows `J0` and `JMAX` of the matrix `A`,

β) for `K=I+1,...,`N_b do:

i) for `J=J0,...,`N_b do:

```
A(K,J)=A(K,J)-A(K,J0)*A(I,J)/A(I,J0),
```

ii) for `J=1,..., I` do:

```
B(K,J)=B(K,J)-A(K,J0)*B(I,J)/A(I,J0),
```

γ) increment `J0`,

Since during the computation, the initial square matrix is transformed into rectangular sub-matrices, the algorithm cannot be simply implemented using an "in place" strategy where the matrix B uses the same storage than matrix A.

REFERENCES.

[1] Frisch U., Hasslacher B., and Pomeau Y., *Phys. Rev. Lett.* 56, 1505 (1986).

[2] An extensive bibliography can be found in the Proceedings of the workshop *Cellular Automata and Modeling of Complex Physical Systems*, Feb. 21 – March 1, 1989, Les Houches, France, P. Manneville, N. Boccara, G.Y. Vichniac, and R. Bidaux editors (Springer–Verlag, 1989) pp. 186–204.

[3] d'Humières D., Lallemand P., and Frisch U., *Europhys. Lett.* 2, 291 (1986).

[4] Wolfram S., *J. Stat. Phys.* 45, 471 (1986).

[5] Frisch U., d'Humières D., Hasslacher B., Lallemand P., Pomeau Y., and Rivet J.–P., *Complex Sys.* 1, 649 (1987).

[6] Zanetti G., *Phys. Rev. A* 40, 1539 (1989).

[7] Cornubert R., d'Humières D., and Levermore D., "A Kudsen layer theory for lattice gases", in the proceedings of the workshop *Lattice Gas Models for PDE's: Theory, Application, and Hardware*, September 6–9, 1989, Los Alamos, USA, to appear in *Physica D* (1990).

[3] The value of the small number must be chosen larger than the round-off errors and not too big, in order to avoid wrong labeling of invariants, our experience indicates that 10^{-3} is a good value for single precision real numbers. If strange results are obtained, the invariants found can be checked *a posteriori* directly from equation (9).

[8] d'Humières D., Qian Y.H., and Lallemand P., "Invariants in Lattice Gas Models", in *Discrete Kinematic Theory, Lattice Gas Dynamics, and Foundations of Hydrodynamics*, R. Monaco editor (World Scientific,1989) pp. 102–113.

[9] Zanetti G., "Counting Hydrodynamic Modes in Lattice Gas Automata Models", in the proceedings of the workshop *Lattice Gas Models for PDE's: Theory, Application, and Hardware*, September 6–9, 1989, Los Alamos, USA, to appear in *Physica D* (1990).

[10] Qian Y.H., "Modèles de Gaz sur Réseaux et Théorie Cinétique sur Réseaux Appliquée à l'Équation de Navier–Stokes", doctoral thesis, Université Paris VI (1990).

[11] Qian Y.H., d'Humières D., and Lallemand P., *C. R. Acad. Sci. Paris II* 308, 585 (1989).

[12] d'Humières D., Lallemand P., and Qian Y.H., "Diffusion Simulations with a Deterministic One–dimensional Lattice–Gas Model", preprint ENS, submitted to *J. Stat. Phys.* (1989).

[13] Hardy J. and Pomeau Y., *J. Math. Phys.* 13, 1042 (1972).

[14] Hardy J., Pomeau Y., and de Pazzis O., *J. Math. Phys.* 14, 1746 (1973).

[15] Hardy J., de Pazzis O., and Pomeau Y., *Phys. Rev. A* 13, 1949 (1976).

[16] S. Chapman and T.G. Cowling, *The Mathematical Theory of Non–Uniform Gases* (Cambridge Univ. Press, 1939).

[17] Hénon M., *Complex Systems* 1, 475 (1987).

[18] Hénon M., *Complex Systems* 1, 763 (1987).

[19] Hénon M., "Optimisation of collision rules in the FCHC lattice gas, and addition of rest particles", in *Discrete Kinematic Theory, Lattice Gas Dynamics, and Foundations of Hydrodynamics*, R. Monaco editor (World Scientific,1989) pp. 146–159.

[20] d'Humières D. and Lallemand P., *Complex Sys.* 1, 599 (1987).

[21] Hadamard J., "Résolution d'une Question Relative aux Déterminants", *Bull. Sci. Math. Ser. 2*, 17, 240 (1893).

[22] Beauchamp K.G., *Walsh Functions and Their Applications*, (Academic Press, 1975).

[23] Press W.H., Flannery B.P., Teukolsky S.A., and Vetterling, *Numerical Recipes: The Art of Scientific Computing*, (Cambridge University Press, 1986).

[24] Margolus N., Toffoli T., and Vichniac G., *Phys. Rev. Lett.* 56, 1694 (1986).

[25] Vichniac G., *Physica D* 10, 96 (1984).

[26] Pomeau Y., *J. Phys. A* 17, L415 (1984).

Applications of Large Neuron Networks

Thadeu Josino Pereira Penna and Paulo Murilo Castro de Oliveira
Instituto de Física, Universidade Federal Fluminense
C.P. 100296 Niterói RJ, Brazil 24000

ABSTRACT

Most recent works in neural networks treat the Hopfield model using Hebb's rule for the synaptic interconnections. In this model the interaction between two neurons is symmetric, allowing an analytic Hamiltonian formalism. Moreover, each neuron interacts with all others and this long range interaction can be treated by mean field approaches. Many interesting results have been obtained from those works.

In real biological neuron systems, however, the interaction between neurons are neither symmetric nor completely uniform (there are many experimental evidences that groups of interacting neurons are linked in complex hierarchical networks). Thus, numerical simulations are the best way to extract informations from more realistic models. The nonlinear behavior and complexity of these systems cannot be modeled in small assemblies of formal neurons: simulations must be performed for large numbers of interacting neurons. In this work we review how to do these computer simulations and show some simple applications.

Work partially supported by brazilian agencies CAPES, FINEP and CNPq.

I) INTRODUCTION

The idea of formal neurons was introduced by Mc Culloch and Pitts in 1943. Each formal neuron can be in two states, active or non-active. A variable S_i assume the values -1 or $+1$ for active or non-active neurons respectively (see figure 1).

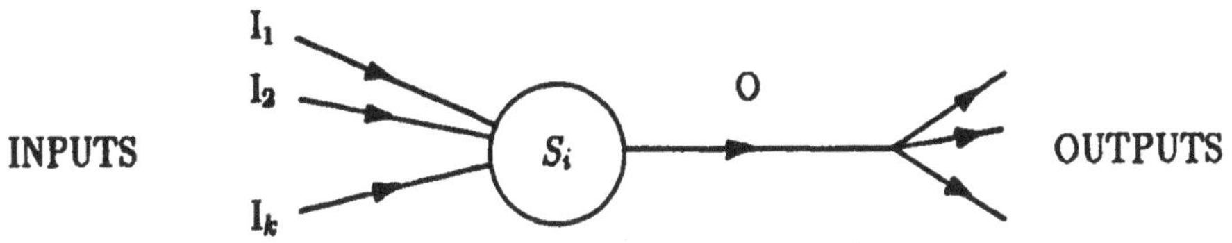

figure 1

The value of S_i depends on the sum of inputs received by neuron i from other neurons $1, 2 \ldots k$ through the corresponding synapses. If this sum surpasses a certain threshold, neuron i becomes active, that is:

$$S_i = -sign \sum_j I_j,$$

where the threshold is taken to be zero. The output released by neuron i in order to influence other neurons has the same sign as S_i, and must be multiplied by the corresponding synaptic efficience. This recipe mimics the electrochemical interactions between real neurons.

An interconnected assembly of many such formal neurons, as shown in figure 2, present a very high degree of nonlinearity, because its collective behavior is qualitatively different from that of the sum of its individual components. The neurophysiologist D. O. Hebb [1)] proposed that the capacity of a neural system to perform tasks (recognition of memorized patterns) resides on the self-organization of the interconnection strenghts, due to external stimuli. Following Hebb's ideas, the stronger the tendency of two neurons to stay in the same state (both active or both non-active), the stronger the synaptic efficience between them. The main point is that memory resides on the synaptic strenght distribution.

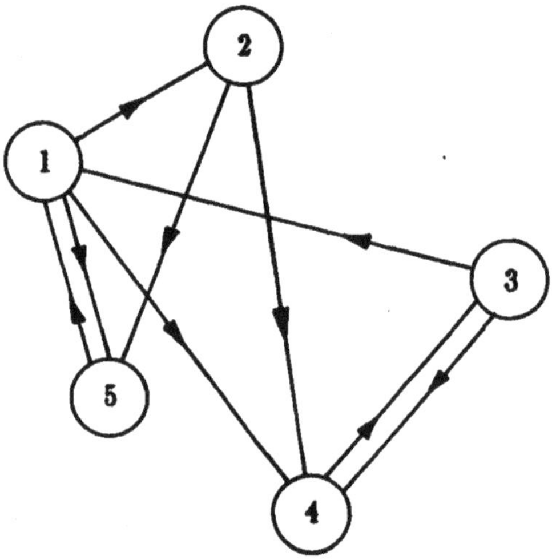

figure 2

A spin-glass like mathematical formulation of these ideas is due to J. J. Hopfield [2]. An energy-like function

$$E = -\sum_{i,j} J_{ij} S_i S_j$$

is defined for a N-neuron network, where $i, j = 1, 2 \ldots N$. The dynamic evolution of the system, described by the equation

$$S_i(t+1) = sign \sum_j J_{ij} S_j(t),$$

minimizes the energy until a steady-state is reached. These steady-states are the patterns memorized by the network. Because slightly different inputs stimuli $\{S_i(t=0)\}$ converge to the same final steady-state, the system is able to recognize a memorized pattern from not-yet-learned inputs. All possible points in the phase space that belong to the basin of attraction of a certain memorized pattern will be recognized by the system.

The fundamental question that remains to be answered is: how to choose the synaptic strenght distribution $\{J_{ij}\}$ in order to memorize just the desired patterns? The most widespread answer to this question is based on Hebb's ideas. Suppose you want to memorize just one pattern

$$\xi^{(1)} \equiv \{\xi_i^{(1)} = \pm 1\}$$

that is a well defined sequence of N values -1 or $+1$. The synaptic strenght distribution

$$J_{ij} = \xi_i^{(1)} \xi_j^{(1)}$$

is adequate, because $\xi^{(1)}$ is just the state for which E is minimum. There is another minimum for the symmetric state $-\xi^{(1)}$ (the negative image), but it contains the same information as $\xi^{(1)}$. The memorized pattern $\xi^{(1)}$ (or $-\xi^{(1)}$) is always recognized, independent of the input stimulus. In order to memorize two patterns $\xi^{(1)}$ and $\xi^{(2)}$ the distribution

$$J_{ij} = \xi_i^{(1)} \xi_j^{(1)} + \xi_i^{(2)} \xi_j^{(2)}$$

can be, but not always, adequate. This learning-by-superposition rule works only if the two memorized patterns are not similar, i.e. if the correlation

$$c(1,2) \equiv \frac{1}{N} \sum_i \xi_i^{(1)} \xi_i^{(2)}$$

between them are small ($\ll 1/\sqrt{N}$). If this condition is not fulfiled neither $\xi^{(1)}$ nor $\xi^{(2)}$ lead to a minimum of E, the real minimum is reached for an intermediate state and one does not succeed in distinguishing the two patterns. For P memorized patterns, the learning-by-superposition algorithm leads to the distribution

$$J_{ij} = \sum_{\mu} \xi_i^{(\mu)} \xi_j^{(\mu)}.$$

The network will work as a content addressable memory device if the following conditions are fulfiled:

$$| c(\mu, \nu) | \ll 1/\sqrt{N}$$

$$\alpha \equiv P/N < \alpha_c = 0.14 \text{ [3]}.$$

II) WHY LARGE N ?

The learning-by-superposition rule has two main weaknesses: the symmetry $J_{ij} = J_{ji}$ of the synaptic strenght distribution; and the fact that all neurons are considered uniformly on the same foot. Real biological neuron systems present asymmetric interconnections as well as groups of neurons specialized in certain tasks. The links between neurons inside the same group are not equivalent to links between neurons of different groups. As analytical mean field approaches are based just in these two unrealistic characteristics, more realistic models can be studied almost exclusively by computer simulations.

An old example of non-symmetric and non-uniform neuron networks is the so-called layered structure shown in figure 3. In this case, the states of the neurons of each layer are defined only by the states of the neurons of the previous layer. The inclusion of hidden neurons improve the performance of the network [4]. This structure has been successfully used in the so-called NETtalk [5], a software able to read loudly english texts with the correct pronounciation. First, in a learning step, the synaptic strenght distribution is self–consistently determined by correctly tuning the pronounciation of an entire page. After this first learning step, other "unknown" pages can be correctly read.

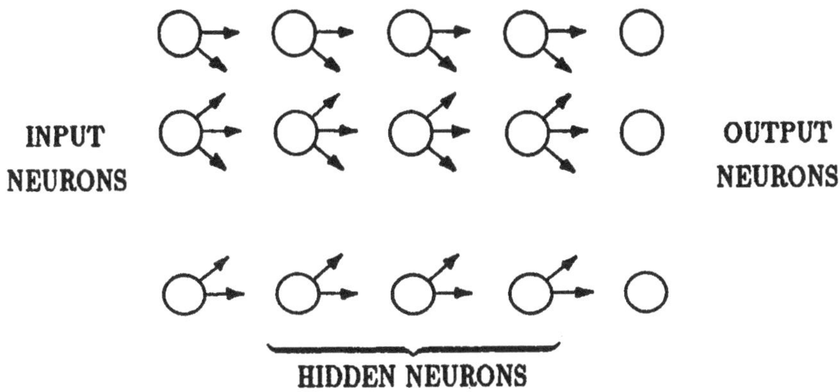

figure 3

Another more recent example is the Gardner's rule of learning

$$J_{ij} = \sum_{\mu} \epsilon_i^{(\mu)} \xi_i^{(\mu)} \xi_j^{(\mu)}$$

where $\epsilon_i^{(\mu)}$ are determined by a well defined optimization process [6]. The factor $\epsilon_i^{(\mu)}$, depending only on i, leads to an asymmetric synaptic strenght distribution.

III) "MULTIPLE NEURON CODE" [7)]

In order to store the state of the whole network bit-by-bit in the computer memory, we introduce the following transformations:

$$\xi_i^{(\mu)} \longrightarrow \varsigma_i^{(\mu)}$$

$$S_i \longrightarrow \sigma_i$$

$$-1 \longrightarrow 1$$

$$+1 \longrightarrow 0$$

The dynamic evolution of the traditional learning-by-superposition Hopfield model

$$S_i(t+1) = sign\Big[\sum_{j=1}^{N} \overbrace{\sum_{\mu=1}^{P} \xi_i^{(\mu)} \xi_j^{(\mu)}}^{J_{ij}} S_j(t)\Big]$$

is now expressed as

$$\sigma_i(t+1) = \frac{1}{NP}\Big[2\sum_{j=1}^{N}\sum_{\mu=1}^{P} \varsigma_i^{(\mu)} \otimes \varsigma_j^{(\mu)} \otimes \sigma_j(t)\Big]$$

where $\otimes$ represents the logical XOR operator and the division by NP is of integer type. In the traditional scheme, one first calculates the $N \times N$ integer matrix J_{ij} and stores it. After this first step, during the dynamical evolution, the current state $\{S_i(t)\}$ of the network is stored in a N-dimensional integer array. In our scheme, we don't store the matrix J_{ij}, but only $\varsigma_i^{(\mu)}$ in P $(N/16)$-dimensional integer arrays (in our 16-bit-word microcomputers). During the dynamical evolution, $\sigma_i^{(\mu)}$ are stored in another $(N/16)$-dimensional integer array.

Besides its memory-saving advantages, the more important feature of this approach is its time-saving characteristic obtained by using rapid logical operations instead of slow algebraic ones. Furthermore, logical operations treats 16 neurons in parallel. The net effect gives a factor far beyond 16 in computer-time saving.

Our trick consists of an interchange of the sums, as explained below. Operations are performed from the right to the left.

$$\sigma_i(t+1) = \frac{1}{NP}[2\sum_{\mu=1}^{P}\sum_{j=1}^{N}\varsigma_i^{(\mu)} \otimes \varsigma_j^{(\mu)} \otimes \sigma_j(t)]$$

only remainder algebraic operation

bit counting operation

invert bits (NOT) *if* $\varsigma_i^{(\mu)} = 1$

only one XOR *operation for* 16 *neurons simultaneously*

Using this approach we succeed in performing simulations for a $N = 10^4$ neuron network with $N^2 = 10^8$ synapses on an IBM–PC running at 4.77MHz (the original 1981 PC), in $\sim$ 10min/stored pattern. On another PC based on the (also old) processor INTEL–80286 running at 12MHz, the time was $\sim$ 3min/stored pattern. For comparison, the largest simulation we know was performed by Amit et al.[3)] on main–frames for $N = 3000$. Furthermore, our $N = 10^4$ network is yet far below the capacity of our PC.

Asymmetric and non–uniform rules can be easily implemented. For instance, the Gardner factor $\epsilon_i^{(\mu)}$ can be simply included before (at right of) the μ summation.

IV) RETRIEVING CORRELATED PATTERNS [7)]

If the correlation

$$x = c(1,2) = \frac{1}{N}\sum_i \xi_i^{(1)}\xi_i^{(2)} = 1 - \frac{2}{N}\sum_i \varsigma_i^{(1)} \otimes \varsigma_i^{(2)} > 0$$

between patterns $\xi^{(1)}$ and $\xi^{(2)}$ is large, i.e. $\gg 1/\sqrt{N}$, the traditional Hopfield model does not work. A simple solution is to divide conveniently the network into small independent pieces (subnets), as explained below.

In this case we have $N(1+x)/2$ "equal" and $N(1-x)/2$ "different" neurons in $\xi^{(1)}$ and $\xi^{(2)}$, as shown in figure 4.

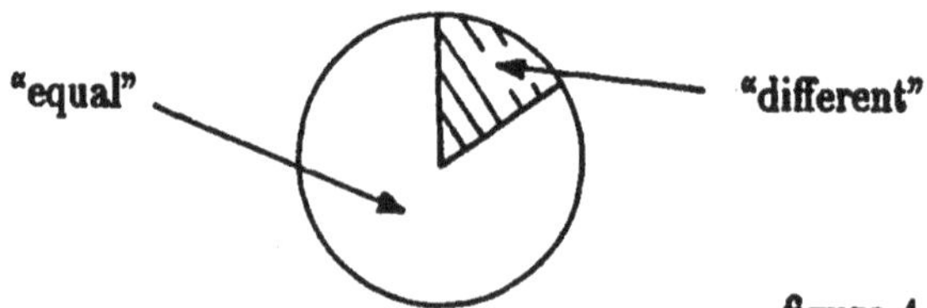

figure 4

The first subnet is formed by any Nx neurons taken from the set of "equal" ones. In this subnet $\xi^{(1)}$ and $\xi^{(2)}$ are equal and can be considered as the same pattern. The other subnet is formed by the $N(1-x)$ remainder neurons, half of which are "equal" and the other half are "different". In this subnet the correlation between $\xi^{(1)}$ and $\xi^{(2)}$ is exactly zero, and thus the traditional Hopfield approach works (see figure 5).

figure 5

For more than two correlated patterns the same method can be applied by a convenient choice of the Nx neurons forming the first subnet or by dividing the subnets into smaller sub–subnets. Clearly, the total number of neurons must be large enough.

V) RETRIEVING LOW–ACTIVITY PATTERNS [8]

The activity

$$a(\mu) \equiv \frac{1}{N}\sum_i \varsigma_i^{(\mu)}$$

of a pattern $\varsigma^{(\mu)}$ is its fraction of active neurons. The traditional Hopfield model works only if the patterns have $\sim 50\%$ activity each. Neither real biological neuron systems nor practical applications present such a high level of activity, in general. A digitalized picture, for instance, has much more white than black pixels, as in this page.

We propose the following recipe in order to overcome this difficulty:

a) take another N–neuron "ghost" network exactly equal to the first one;

b) include noise into the patterns, by turning originally white pixels ($\varsigma_i = 0$)

into black ($\varsigma_i = 1$), on one or other N–network, but *never the same pixel on both*, until $a(\mu) = 50\%$ for all patterns;

c) join both network forming a $2N$–net with $4N^2$ synapses;

d) use the traditional Hopfield dynamics for the $2N$–network;

e) after convergence, eliminate noise through an AND operation between the two N–nets.

We have tested this recipe for patterns with 30%, 10% and 5% of activity. The retrieval was complete until the saturation fraction of stored patterns $\alpha_c = 0.28$ (this value is twice the normal one because we used twice the number of neurons). Above α_c, in contrast with traditional Hopfield model whose performance worses for decreasing average activity, we found better retrieval efficience for smaller average activity, indicating that errors occur mainly in the noise included into the patterns.

ACKNOWLEDGEMENTS

We are indebted to our colleagues Evaldo M. F. Curado and Francisco A. Tamarit for helpful discussions about this subject, and Oziel de Alcantara Bonfim for a critical reading of the manuscript.

REFERENCES

1) D. O. Hebb, *The Organization of Behavior*, Wiley N.Y. (1949).

2) J. J. Hopfield, *Proc. Nat. Acad. Sci. USA* **79**, 2554 (1982).

3) D. J. Amit, H. Gutfreund and H. Sompolinsky, *Ann. Phys. NY* **173**, 30 (1987).

4) T. J. Sejnowski, P. K. Kienker and G. E. Hinton, *Physica* **D22**, 260 (1986).

5) T. J. Sejnowski and C. R. Rosenberg, *Technical Reports JHV/EECS - 86/01 John Hopkins University*, (june 1986).

6) see the Elizabeth Gardner memorial issue of *J. Phys.* **A**, (june 1989).

7) T. J. P. Penna and P. M. C. de Oliveira, *J. Phys.* **A**, to appear (1989).

8) T. J. P. Penna and P. M. C. de Oliveira, *preprint*, (1989).

LANGEVIN DESCRIPTION OF ROTATING MOLECULES

R. W. Gerling

Institut für Theoretische Physik I der Universität Erlangen-Nürnberg
Glückstraße 6, D-8520 Erlangen, Fed. Rep. of Germany
and
Center for Simulational Physics, University of Georgia,
Athens, GA 30602, USA

ABSTRACT

The Langevin equation is a powerful tool to study the rotational motion of molecules in gases, liquids and solids. The method is described and applied to the rotational motion of the NH_4 tetrahedron in the molecular crystal NH_4Cl. We also present a simple model to study translational-rotational coupling.

1. THE LANGEVIN EQUATION

The Langevin model has been quite successfully used to study rotational motion of molecules in gases, liquids, and solids[1-9].It is an excellent tool to study reorientational motion of molecules and molecular groups in molecular crystals.

The Langevin equation, which was presented the first time in 1908 by Paul Langevin[10], is given by

$$\ddot{\vec{x}} = -\frac{1}{m}\nabla V(\vec{x}) - |\dot{x}|^2\vec{x} - \Gamma\dot{\vec{x}} + G\frac{\vec{F}(t)}{m}. \qquad (1)$$

The acceleration $\ddot{\vec{x}}$ of a particle is given by a force term resulting from a rotational potential $V(\vec{x})$, a centripedal term, a friction term $-\Gamma\dot{\vec{x}}$ and a stochastic force $G\vec{F}(t)$. The strength of the stochastic force is G and $\vec{F}(t)$ is a fluctuating function of time which is characterized by

$$
\begin{aligned}
< \vec{F}(t) > &= 0 \\
< \vec{F}(t) \cdot \vec{F}(t') > &= \delta(t - t').
\end{aligned} \tag{2}
$$

Normally the Langevin model is applied to a heavy particle in a bath of light particles. This means that the velocity of the heavy particle is small compared with the velocities of the bath particles. Thus the separation of time scales is justified.

In the case of rotational motion the situation is quite different. The particle of interest is of the same size (often it is exactly the same molecule) as the bath particles. Hence the conventional premise of a heavy particle in a bath of light ones is not correct. The interaction energy of two molecules m and n with orientations denoted collectively by Ω_m and Ω_n and a center of mass distance $\vec{r}_{mn}$ is usually[11] split into its isotropic and anisotropic parts

$$
V(r_{mn}, \Omega_m, \Omega_n) = V_i(r_{mn}) + V_a(r_{mn}, \Omega_m, \Omega_i). \tag{3}
$$

By definition V_i is the spherical average of V, and therefore $V_i =< V >$ and $< V_a >= 0$. For globular molecules the anisotropic part V_a of the potential is much weaker than the isotropic part. This is clearly demonstrated by the existence of ODIC phases, where the central forces have established a crystalline order of the centers of mass of the molecules, but the weak anisotropic forces cannot achieve an orientational order of the molecules. In dense phases of globular molecules the rotational agitation of an individual molecule is permanent but weak at the same time. Therefore the angular momentum of a molecule can change only little on the time scale of the fluctuations of the torques created by its neighbors. Thus the separation of time scales leading to the Langevin picture is justified also for rotating molecules.

To integrate the Langevin equation (1) we used a very simple and straightforward difference scheme

$$
\dot{x}_i^{n+1} = \ddot{x}_i^n \Delta + \dot{x}_i^n \tag{4a}
$$

$$
x_i^{n+1} = \frac{1}{2}\ddot{x}_i^n \Delta^2 + \dot{x}_i^n \Delta + x_i^n. \tag{4b}
$$

n labels the integration step and Δ is the length of the time interval in the simulation. This technique is correct up to order Δ^2.

After every time step the conditions $|\vec{x}| = 1$ and $\vec{x} \cdot \dot{\vec{x}} = 0$ must be satisfied to keep the point on the sphere. Therefore the length of the radius-vector and the direction of the velocity are rescaled properly. The stochastic force is modeled by a vector $\vec{F}_0(t)$, whose components are Gaussian distributed random numbers at any time t. $\vec{F}(t)$ is the projection of $\vec{F}_0(t)$ on a plane tangential to the sphere at the point $\vec{x}(t)$. In a Gauss-Markov process the probability that after time t a collision has occurred is given by[12]

$$p(t) = \frac{1}{T} \exp(-\frac{t}{T}), \tag{5}$$

where T is the mean time between two collisions. The probability that a collision has happened after a time Δ is $w(\Delta) \approx \frac{\Delta}{T}$. We have chosen a random torque of the following form:

$$F_i(t) = \sum_n f(t - n\Delta)\eta_i(n)\Theta(w - a_i(n)), \tag{6}$$

where the sum over n runs over all time steps. The $\eta_i(n)$ are Gaussian distributed random numbers; the $a_i(n)$ are equally distributed random numbers from the interval $[0, 1]$. $\Theta(x)$ is Heaviside's step function. The function f(x) is unity for arguments between zero and Δ and zero elsewhere. We generated a pair of independent Gaussian distributed random numbers with the method of Box and Muller[13]. The prefactor of the stochastic force G is given by the fluctuation-dissipation theorem

$$G = \sqrt{\frac{2\Gamma T}{\beta \Delta^2}}. \tag{7}$$

β is the inverse temperature, measured in units of k_B. The factor $\frac{T}{\Delta^2}$ has its origin in the finite time steps in the computer simulation, leading to a torque which is not strictly δ-correlated.

2. NH_4Cl

In this example we want to study the rotational motion of the NH_4 tetrahedron in an orientational potential, which stems from its interaction with the Cl-neighbors in the molecular crystal NH_4Cl.

Hüller and Kane[14] used an electrostatic model to calculate the orientational potential of the ammonium ion in the cubic field of the surrounding halide

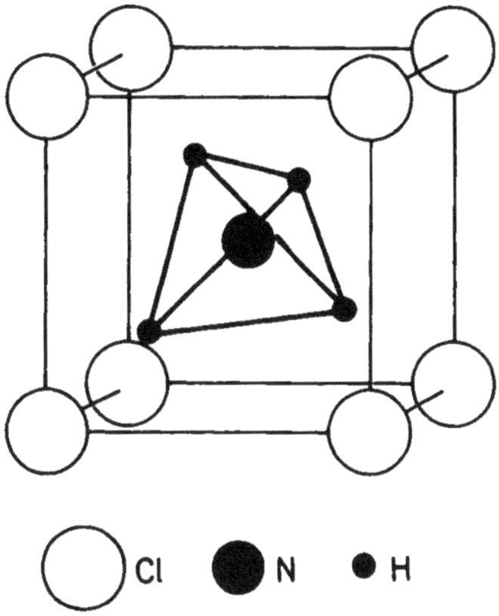

Figure 1: *The lattice structure of NH_4Cl. The Cl^- ions are arranged in a cubic structure. The figure shows one of the two possible equilibrium orientations.*

ions and in the field of the surrounding ammonium ions. The structure of NH_4Cl can be seen in Fig. 1. The rotational potential of the ammonium ion is expanded in terms of symmetry adapted functions, in this case in terms of the cubic rotator functions $H^{(\ell)}_{\mu\nu}(\xi,\eta,\zeta)$ [15]. To avoid numerical problems integrating the differential equation quaternions $\vec{\tau} = (\tau_1,\tau_2,\tau_3,\tau_4)$[5,16] instead of Euler angles (ξ,η,ζ) were used. The rotational potential now reads

$$V_{rot}(\vec{\tau}) = B^3_{11}H^{(3)}_{11}(\vec{\tau}) + B^4_{11}H^{(4)}_{11}(\vec{\tau}) + B^6_{11}H^{(6)}_{11}(\vec{\tau}). \tag{8}$$

The cubic rotator function $H^{(\ell)}_{11}(\vec{\tau})$ is a polynomial in the τ_i of order 2ℓ. The coefficients B^{ℓ}_{11} can be found in Table 1.

The simulation of the Langevin equation containing the potential (8) can be done by using the difference scheme (4) identifying x_i with τ_i; the equation describes the motion of a point on the four-dimensional unit-sphere in the potential (8) driven by a stochastic force and hindered by a friction term. For the stochastic force the form (6) was used. The friction constant Γ was varied from 0.25 to 3.75. The temperature was varied between about 287 Kelvin and 1575 Kelvin. For each data point 10 runs of $2.5 \cdot 10^5$ time steps were averaged.

coefficient	ordered	disordered
B^3_{11}	-323.20 K	-
B^4_{11}	-2221.38 K	-2213.90 K
B^6_{11}	-367.81 K	-354.28 K

Table 1: *The coefficients B^ℓ_{11} for the ordered and disordered phase of NH_4Cl. The term with tetrahedreal symmetry ($\ell = 3$) is not present in the disordered phase. From Ref. 14.*

Töpler et al.[17] were able to measure the jump-rates ν_i of the ammonium tetrahedron around its four-fold and three-fold symmetry-axes separately. Their finding was that $\nu_4 \approx \frac{2}{3}\nu_3$. From the static potential (8) one can estimate that $\nu_4 \approx 60\nu_3$.

The results from the simulation allow us to follow the path of the tetrahedron, so one can easily determine the jump rate around the different symmetry-axes by analysing the motion. Figure 2 shows typical results for the motion. The distance from a given minimum is plotted as a function of time. The time scale is in units of 10Δ. The number above the abscissa denotes the minimum. If the number changes the particle has jumped from one minimum to another. Figure 2a shows a result for small friction; the motion is nearly oscillatory. Figure 2b shows results for a ten times larger friction: the motion is stochastic.

In Figure 2c the key result can be seen: the tetrahedron, once jumping, does not jump just to the next minimum, but it makes multiple jumps through several minima. When the particle has picked up enough energy to cross a potential barrier, it does not lose its energy until it has passed through several other minima. A jump is finished when the total energy of a particle is less than the lowest barrier (barrier for 90°-jumps), because in this case the particle cannot leave the minimum without any further agitation from the stochastic force. The percentage of multiple jumps can be changed almost continously by varying the friction parameter. Figure 3 shows the fraction of single jumps as a function of the friction parameter.

If a given jump is a single jump, the probabilities that a jump is a 90° jump (p_s^{90}) or a 120° jump (p_s^{120}) are almost independent of the friction. The same

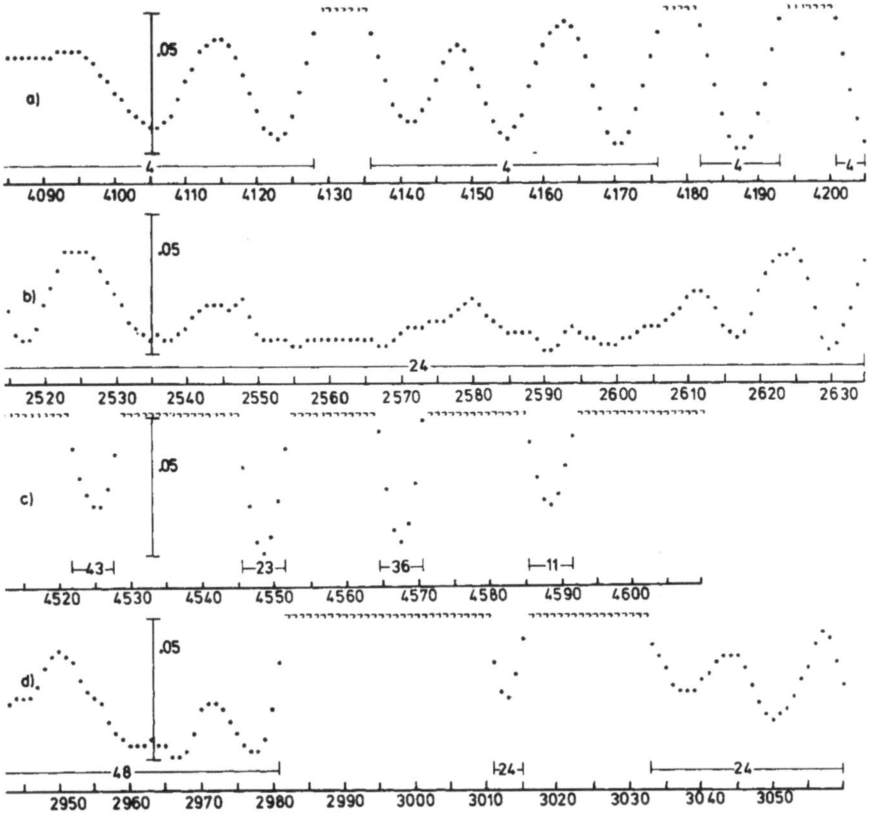

Figure 2: *The figure shows the square of the distance of the tetrahedron from the minimum. The abscissa shows the time in units of* 10Δ*; the number above the abscissa denotes the minimum.(a) The motion is oscillatory and* $\Gamma = 0.25$*. (b) The friction parameter* $\Gamma = 2.50$ *and the motion is stochastic. (c) Now the figure shows a multiple jump. All three jumps are* 90° *jumps around a (100) axis. The friction* $\Gamma = 0.25$*. (d) This figure shows a single jump. The jump is a* 90° *jump around a (100) axis.* $\Gamma = 2.50$.

is true for multiple jumps. We found from the simulation

$$\begin{aligned} p_s^{90} &= 0.88 \qquad & p_s^{120} &= 0.13 \\ p_m^{90} &= 0.29 \qquad & p_m^{120} &= 0.40. \end{aligned} \tag{9}$$

Mixing single and multiple jumps we can produce the experimental results.

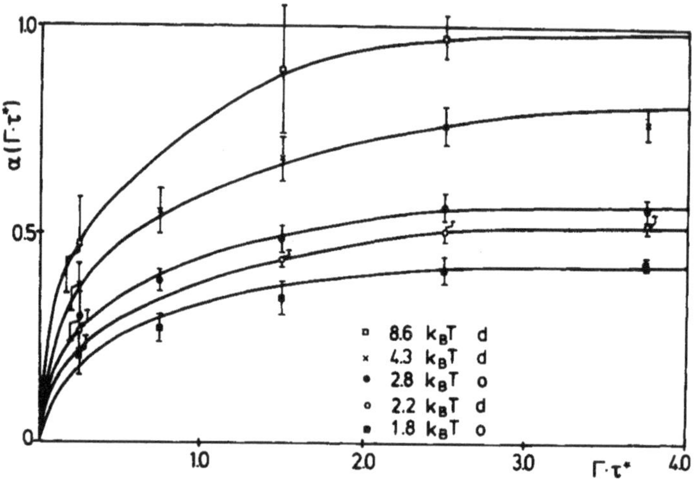

Figure 3: *The figure shows the proportion α of single jumps as a function of the friction parameter Γ for different barrier heights as indicated.*

The proportion of single jumps α can be determined by

$$\begin{aligned} 0.88\alpha + 0.29(1-\alpha) &= \nu_4 = \frac{2}{3}\nu_3 \\ 0.13\alpha + 0.40(1-\alpha) &= \nu_3. \end{aligned} \tag{10}$$

The solution is $\alpha = 0.40$. Thus we have been able to explain the experiment by the dynamical effect of multiple jumps. The neutron scattering experiment cannot see the difference between multiple jumps and single jumps, because the jump duration in both cases is short compared to the residence time in a minimum. As shown in Fig. 4 a 120° jump arround a threefold axis is equivalent to two 90° degree jumps arround to perpendicular fourfold axes. A neutron scattering experiement cannot observe the difference between these two events. Hence the event shown in the upper part of Fig. 4 is counted as a normal 120° degree jump, thus increasing the proportion of 120° jumps.

Figure 5 shows a comparison of the quasi-elastic line-width with a first order approximation extracted from the computer simulation. The correlation function $< \vec{x}(t)\cdot\vec{x}(0) >$ was fitted with an exponential function for times larger than the inverse friction parameter. The decay-constant was used as the estimate for

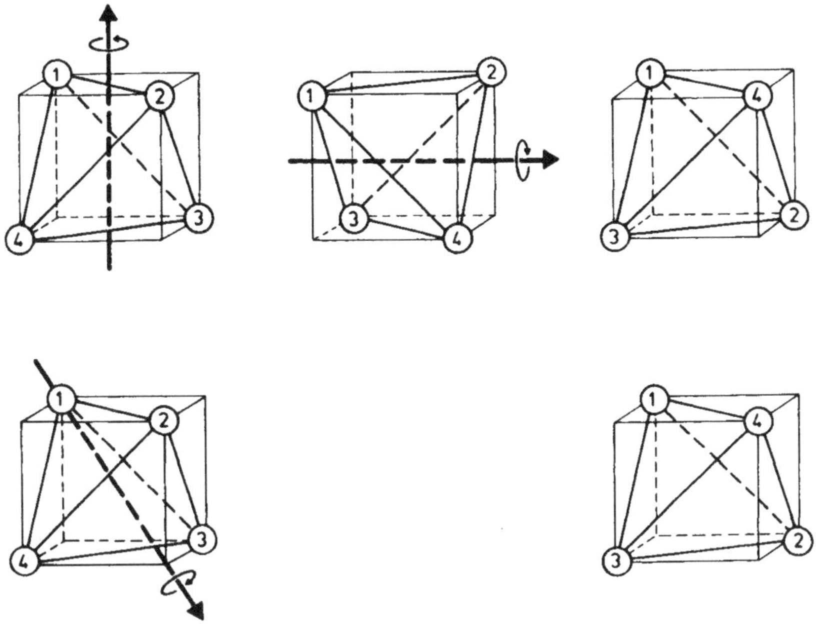

Figure 4: *A 120° jump arround a threefold axis is equivalent to two 90° degree jumps arround to perpendicular fourfold axes. A neutron scattering experiement cannot observe the difference between these two events.*

the width. The constants follow an Arrhenius law, which allowed us to extrapolate into the experimental temperature range. This simple estimate does not include a wave-vector dependence. The activation energies from the the Arrhenius law were

$$E_{ordered} = 214meV, \qquad E_{disordered} = 181meV, \tag{11}$$

which agrees very well with the values of Hüller and Kane[14] and Töpler et al.[17].

3. TRANSLATIONAL ROTATIONAL COUPLING

We have simulated a simple planar model consisting of four harmonic oscillators moving on the four sides of a square with a free dumbbell-rotator in the center of the square. The four oscillators are coupled to a heat bath by

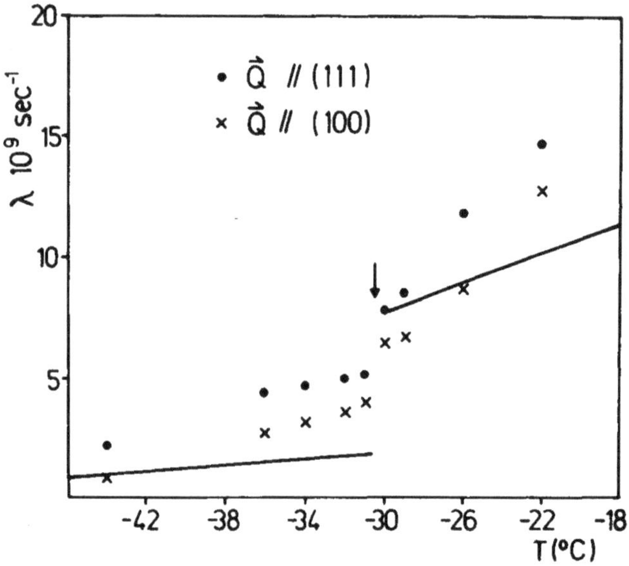

Figure 5: *The width λ of the Lorentzians as a function of temperature. The points are experimental data from Ref. 17. The lines are from the numerical simulation. The arrow indicates the phase transition.*

a stochastic force and a linear friction. Fig. 6 outlines the geometrical relations. The four oscillators interact with the dumbbell in the center via a simple potential. One can think of several different interactions between the central particle and the outer particles. We chose an attractive $\frac{1}{r}$ potential. The two parameters of the model are the length of the rotator $2a$ and the length of a side of the square $2b$. The potential is given by

$$\begin{aligned} V(x_1,x_2,x_3,x_4,\varphi) = -k\Big(&f_1(x_1,\varphi)+f_1(x_3,\varphi)+f_2(x_1,\varphi)+f_2(x_3,\varphi)\\ &f_3(x_2,\varphi)+f_3(x_4,\varphi)+f_4(x_4,\varphi)+f_4(x_4,\varphi)\Big), \end{aligned} \tag{12}$$

where the x_i $(i=1,2,3,4)$ are the displacements of the oscillators measured from the middle of the four sides in a counter-clockwise direction and φ is the rotational angle of the dumbbell rotator. For simplicity we have introduced the four functions

$$f_1(x,\varphi) = \Big(a^2+b^2+x^2-2a(b\cos\varphi + x\sin\varphi)\Big)^{-1/2} \tag{13a}$$

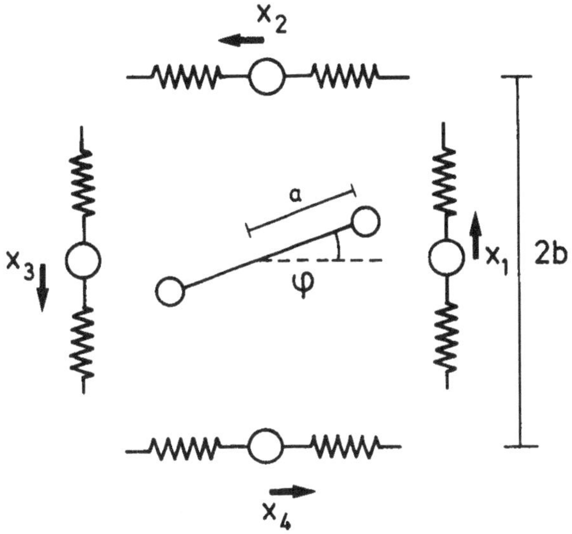

Figure 6: *The figure shows the geometrical outline of the system. The masses indicated by the spheres were treated as point-masses.*

$$f_2(x,\varphi) = \left(a^2 + b^2 + x^2 + 2a(b\cos\varphi + x\sin\varphi)\right)^{-1/2} \tag{13b}$$

$$f_3(x,\varphi) = \left(a^2 + b^2 + x^2 - 2a(b\sin\varphi - x\cos\varphi)\right)^{-1/2} \tag{13c}$$

$$f_4(x,\varphi) = \left(a^2 + b^2 + x^2 + 2a(b\sin\varphi - x\cos\varphi)\right)^{-1/2}. \tag{13d}$$

The equations of motion for the system are

$$\ddot{x}_i = -\Omega^2 x_i - \frac{1}{m}\nabla_{x_i} V(x_i,\varphi) - \Gamma\dot{x}_i + G^{1/2}\frac{F_i(t)}{m}, i = 1\dots 4 \tag{14a}$$

$$\ddot{\varphi} = -\frac{1}{I}\nabla_{\varphi} V(x_i,\varphi), \tag{14b}$$

where m and I are the mass of the identical oscillators and the moment of inertia of the rotator, respectively. The equations were integrated by using the difference scheme (4). The stochastic force pulses were modeled by the form given in (6).

The static potential for the rotator is shown in the bottom part of Fig. 7.

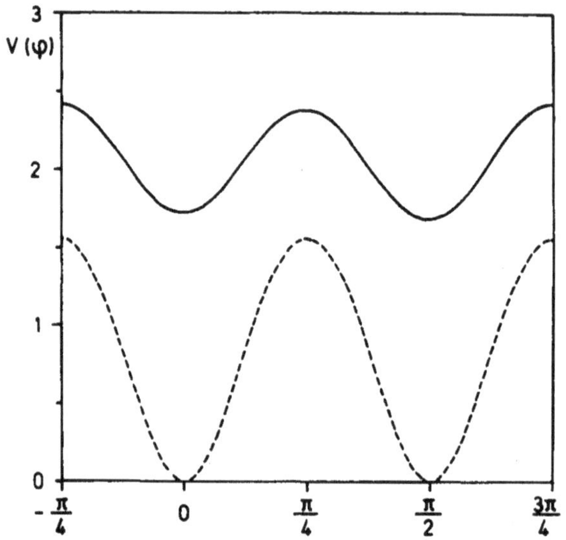

Figure 7: *The figure shows the static potential in the lower part and the time averaged potential in the upper part.*

In the upper part of Fig. 7 the time-average of the potential is shown:

$$< V(x_i(t), \varphi) >_t = \frac{1}{M} \sum_{n=1}^{M} V(x_i^n, \varphi). \tag{15}$$

Due to the symmetry of our model the potential is always π-periodic with respect to φ. The rotator will in most cases be observed around the minima at $\varphi = n \cdot \frac{\pi}{4}, n \in N$. To observe its behavior best we will from now on fold φ in the interval $]\frac{-\pi}{4}, \frac{3\pi}{4}]$. Our investigations[18] show that the effects we want to study are most obviously observed if we use $b = 2a$.

For these values of the parameters the height of the potential wall for the static case becomes

$$V_b = V(x_i = 0, \varphi = \frac{\pi}{4}) - V(x_i = 0, \varphi = 0) \approx \frac{k}{a} \cdot 0.3115. \tag{16}$$

We set the energy of the bath to $k_B T = 1$, $a = 1$ and $k = 5$ which results in $V_b \approx \frac{3}{2} k_B T$. The transition-rate for the rotator in the static case can

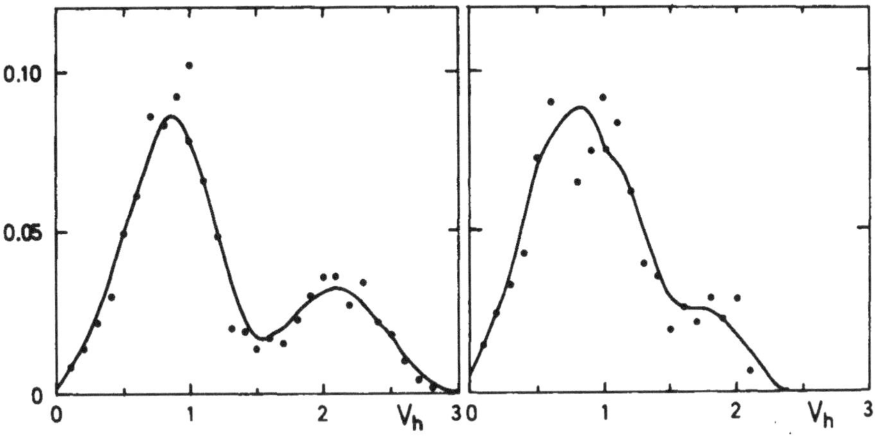

Figure 8: *The figure shows the distribution of the barrier heights. In the left part both barriers are counted, in the right part only the lower of the two barriers is taken into account.*

be approximated by the probability of finding it at the maximum of the potential. That is equal to $\exp(\frac{-V_b}{k_B T}) \approx \exp(-1.5) \approx 0.22$ which is necessary for numerical simulations. In reality $\frac{V_b}{k_B T}$ is far larger, resulting in much lower jump-rates. To keep the calculation-times on the computer within reasonable limits this ratio must be smaller than in reality.

The only factor which could change the probability of a transition is a substantially lower mean energy barrier in the dynamical system. This can in fact be observed.

One possibility of studying the dynamics is to look at the distribution of the energy-barriers to be passed by the rotator. Our studies have shown that most of the time the energy-barriers are of different height. The rotator normally crosses the lower energy-barrier. Taking this into account we generated two graphs (Fig. 8) showing the height-distribution of the energy-barrier for the regular system and for the case where the higher potential-wall is neglected. The average barrier-height is in the regular system $< E_1 >= 1.237 k_B T$ and in the other case $< E_2 >= 1.001 k_B T$. Thus we see that the dynamics of the system causes a shift of the mean energy-barrier from $V_b = 1.5 k_B T$ in the static case to $< E_2 >= 1.0 k_B T$ which can be regarded as the mean energy-barrier for the rotator to be crossed.

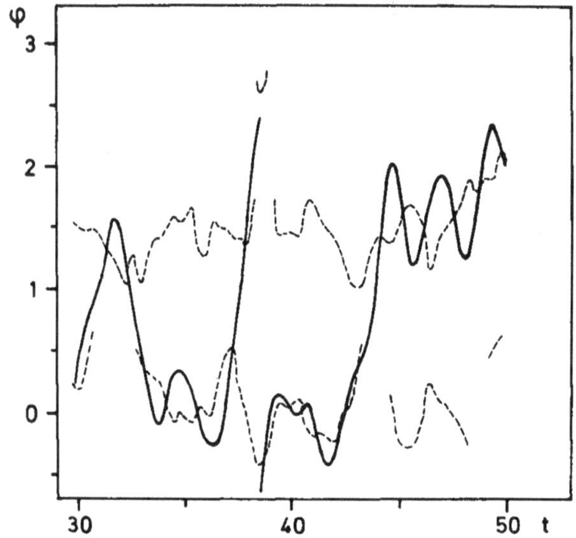

Figure 9: *The figure shows the time evolution of the position of the minima (dashed lines) and the orientation of the rotator (solid line). Note the disappearance of the minima.*

Another access to this problem consists of simply observing the localization of the minima in time. It sometimes occurs that, resulting from the positions of the oscillators, one minimum vanishes and goes over to e.g. a saddlepoint. In this case a transition of the rotator is primarily determined by its environment and not by its kinetic energy.

In Fig. 9 we show the position of the rotator and the positions of the minima. The positions of the minima are moving and the rotator is oscillating around this position. The positions of the two minima oscillate around their static positions $\varphi_1 = 0$ and $\varphi_2 = \pi/2$. Approximately at times $t \approx 31$ and $t \approx 43$ one of the minima is disappearing and the rotator is moving to the other minimum. The transitions are caused by the vanishing of the minimum in which the particle is oscillating.

In order to provide a better insight into the dynamics of one transition Fig. 10 shows a time-sequence of several pictures of the potential including the position of the rotator (marked by the arrow). At time $t = 42.30$ we have two

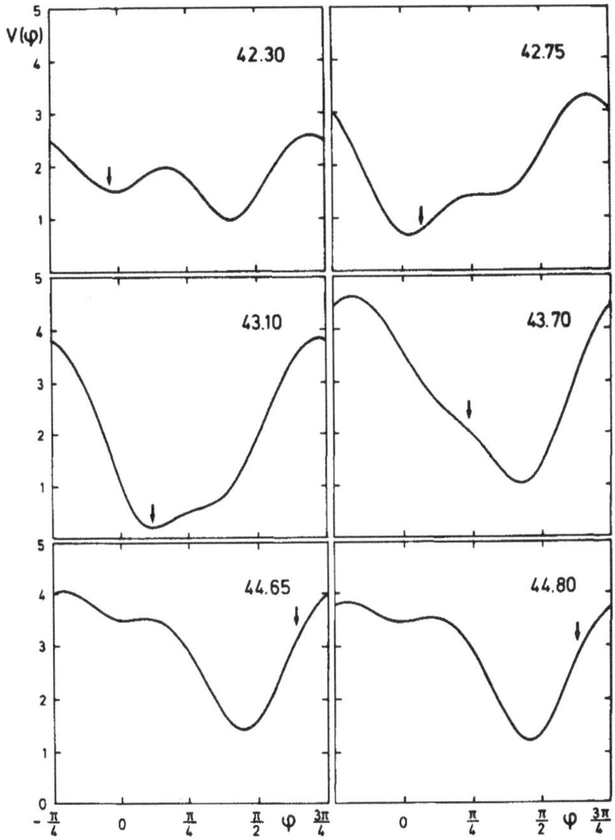

Figure 10: *The figure shows the potential with the orientation of the rotator marked by an arrow for six different times. The times are chosen such that the disappearance of one minimum can be seen clearly.*

distinct minima. At time $t = 43.10$ only one minimum is left. Now the surrounding deforms, moving the particle to the other minimum position (around $\varphi = \pi/2$). After this event the minimum at $\varphi = 0$ reappears. The jump from one minimum to the other minimum has happened without thermal activated barrier crossing. This can be regarded as phonon assisted rotational jumping. Of course the parameters were choosen such that this effect could be demonstrated. In a more real system thermal activated jumps would also be present.

4. CONCLUSIONS

The simulation of the Langevin equation has the advantage that one very easily gets information about the detailed dynamics of the system. Such information is not available from other methods of solving the problem. All other methods calculate directly averaged quantities, e.g. correlation functions, where the numerical Langevin approach probes one particular realisation of the stochastic force. The direct access to the trajectory of the system gives additional insight, making the simulation of the Langevin equation a very powerful tool in understanding the detailed dynamics of a lot of systems, of which the rotational problems presented here are only an example.

5. ACKNOWLEDGEMENTS

I want to thank the Alexander von Humboldt Stiftung for the award of a Feodor-Lynen-fellowship. Helpful discussions and fruitful collaborations with A. Hüller, D. Klotz and A. Rau are acknowledged. The work is partly supported by the National Science Foundation (NSF) Grant No. DMR-86-03605 and by the "Advanced Computational Methods Center" at the University of Georgia.

6. REFERENCES

1) Gerling, R.W., in **Computer Simulation Studies in Condensed Matter Physics II: New Directions** eds. Landau, D.P., Mon, K.K., and Schüttler, H.B., (Springer, Berlin, Heidelberg 1989).
2) McClung, R.E.D., J. Chem. Phys. **73**, 2435 (1980); **75**, 5503 (1981).
3) Levi, G., Marsault, J.P., Marsault-Hérail, F., and McClung, R.E.D., J. Chem Phys. **73**, 2443 (1980).
4) Levy, R.M., Karplus, M., and McCammon, J.A., Chem. Phys. Lett. **65**, 4 (1979).
5) Gerling, R.W. and Hüller, A., Z. Phys. B **40**, 209 (1980).
6) Gerling, R.W., Z. Phys. B **45**, 39 (1981).
7) Gerling, R.W. and de Raedt, B., J. Chem. Phys. **77**, 6263 (1982).
8) Gerling, R.W. and Hüller, A., J. Chem. Phys. **78**, 446 (1983).
9) Klotz, D. and Hüller, A., Z. Phys. B **63**, 69 (1986).
10) Langevin, P., Compt. Rend. **43**, 530 (1908).
11) Silvera, I.F., Rev. Mod. Phys. **52**, 393, (1980).

12) Schneider, T. and Stoll, E., Phys. Rev. B **17**, 1302 (1978).
13) Box, G.E.P. and Muller, M.M., Ann. Math. Stat. **29**, 610 (1958).
14) Hüller, A. and Kane, J.W., J. Chem. Phys. **61**, 3594 (1974).
15) Brot, C. and Darmon, I., Mol. Phys. **21**, 785 (1971).
16) Hund, F., Z. Physik **51**, 1 (1928); Furry, W.H., Phys. Rev. **107**, 7 (1957).
17) Töpler, J.M., Richter, D.R., and Springer, T., J. Chem. Phys. **69**, 3170 (1978).
18) Rau, A. and Gerling, R.W., preprint (1989).

Green's Function Monte Carlo: From Statistical Mechanics to Quantum Mechanics and Back

Peter J. Reynolds

Physics Division
Office of Naval Research
800 North Quincy Street
Arlington, VA 22217 USA

ABSTRACT
Green's function Monte Carlo (or GFMC for short) can be traced back to the early days of the Monte Carlo method. Although GFMC is firmly grounded in many sub-fields of physics, it has probably its closest affinity to statistical mechanics. It is easiest to visualize GFMC as simply a random walk process with branching. Although this suggests its use in many statistical mechanics problems, such as in the simulation of polymers, by a careful choice of rules the random walk provides a stochastic solution to the Schrodinger equation. In this quantum setting, the method has been applied to numerous problems ranging from condensed-matter physics, including much work on liquid helium, to the statistical mechanics of particles in solvents, atomic and molecular physics, plasmas, and nuclear physics. GFMC also has numerous applications in the study of phase transitions where quantum mechanics plays a role. We provide a discussion of GFMC from a pedagogical point of view, mentioning applications where relevant. We focus on the simulation of atoms and molecules, and in particular on the high power-law dependence of the computational cost on atomic number. This scaling turns out to result from a type of critical slowing down. Adapting methods that have been used to "accelerate" statistical mechanics simulations near phase transitions, it is found that GFMC simulations benefit significantly. Other approaches, using pseudopotentials and hybrid GFMC-Metropolis algorithms, are also discussed, and their advantages and disadvantages are weighed.

HISTORICAL BACKGROUND

The use of random numbers to perform simulations of nature only became practical with the advent of computing machinery. Nevertheless, the ideas behind Monte Carlo—at least in a primitive form—had existed in the statistics community from the early 1900's. Using tables of random numbers, statisticians performed "model sampling" of distributions, and also were able to approximate high-dimensional integrals. The well-known Student's t-distribution was first derived by this sort of Monte Carlo. Almost half a century later, and totally independently, physicists and mathematicians in the late 1940's began to realize that difficult mathematical problems could be cleverly recast in terms of stochastic processes. It was the advent of mechanical, and later electronic, computers that made this approach competitive for the first time with analytic methods. In fact, during WWII, particle diffusion problems required in weapons design work were often solved by Monte Carlo. Complicated mathematics was avoided by modeling nature directly.

After the war, a seminal symposium on Monte Carlo methods[1] was held in late June of 1949. There Dr. A. S. Householder defined Monte Carlo as "...an artificial stochastic model of a physical or mathematical model..." The new idea, according to him, was the application to equations arising from *non-probabilistic* contexts. This concept was chiefly due to Ulam and von Neumann (and to a lesser extent to Fermi), who were seeking better methods to obtain answers in light of the newly-invented and revolutionary tool, the computer. At the meeting, however, there was a degree of discord over the definition of Monte Carlo. Should it include all random sampling procedures or only ones used to solve *deterministic* problems? Many preferred a still more restrictive definition, allowing only approaches involving a "swindle." It was realized that one way to make Monte Carlo "trickier" is to use weights. The most obvious way is to have the weights inversely proportional to the factor by which the probability of occurrence has been distorted. This is the essence of importance sampling—which was to play a key role. If a method had no trick, it was disparagingly called "model sampling."

Ironically, despite this desire for cleverness, it was generally the case that the "deterministic" problems being treated involved the re-translation of analytic equations back to their probabilistic roots (e.g. the diffusion of particles). The exception to this was that, in 1947-49, Metropolis and Ulam, following a suggestion of Fermi's, used Monte Carlo to calculate the ground-state energy of the Schrodinger equation.[2] Fermi proposed that one could use a suitable "random walk" process, by noting the similarity of the time-dependent Schrodinger equation to a diffusion equation. Ulam formulated a random walk on a lattice, with "reproduction" possible at each lattice site. Moves were always discrete. Later, Kac also pursued a Monte Carlo treatment of the Schrodinger equation. Although work continued into the 1950's, no improvements were found over more ordinary numerical methods for treating quantum mechanics, and subsequently the approach faded from the scene.

Surprisingly (to me), Monte Carlo back then was already quite sophisticated. Although the famous Metropolis algorithm[3] for sampling efficiently from the Boltzmann distribution (or more generally any known distribution) was yet to come, much was known. For example, techniques for matrix inversion, for the solution of partial differential equations, and for solving Fredholm-type integral equations were already being advanced.[1] The theoretical basis was already there. But important details were missing.

In the 1960's, Kalos developed a far more sophisticated Monte Carlo approach to quantum mechanics,[4-5] namely Green's function Monte Carlo. Still, the seeds of his approach can already be seen in the proceedings of the second Monte Carlo conference, held in 1954. For example, Kalos' iterative procedure for sampling from the exact Green's function resembles the approach described there for solving Fredholm integral equations.[6] Nevertheless, the synthesis was brilliant, and opened the door again to Monte Carlo treatments of quantum mechanics. Kalos' scheme has since branched out into a number of useful variant forms. All embody the use of Green's function propagators, and allow one to work in a continuum space. Kalos further developed an extension to the original approach that allows one to sample the wave function itself[5,7] (in addition to energies). Of particular significance, he also formulated an importance sampling scheme[8] for GFMC that made it numerically practical, and with his co-workers applied it to condensed-matter problems, particularly liquid helium,[9] and nuclear few-body systems.[4-5] This line of work continues to this day, both in Kalos' group and in groups that have spun off from his. In the 1970's and 80's, Ceperley formulated key additions to the theory, including more efficient techniques of sampling the exact Green's function[10] and, more significantly, general ways to handle fermions[11] (which by their statistics cause the wave function to be antisymmetric, and thus lead to problems with positive probability densities). Ceperley also helped formulate a greatly simplified GFMC which makes use of a short-time approximation.[12]

Independently, in the 1970's Anderson discovered that one could solve problems in quantum chemistry by random walks.[13] In this context, he was the first to treat fermion problems with Monte Carlo. His approach, in retrospect, was basically the same as Ceperley's short-time approximation method, but without Kalos' importance sampling. While these two schools merged in the late 1970's, other variants of GFMC were still being born, and were being applied to many classes of problems. In recent years, the nature of the approximations and limitations of the different approaches have been clarified, while its basic appeal has led to a growing following.

GFMC has a clear interpretation in terms of path integrals, providing another avenue for understanding the quantum nature of this Monte Carlo method. During the 1970's, an independent line of theoretical development led to an explicit path-integral Monte Carlo method for dealing with quantum mechanics.[14] This path integral approach and the Green's function method are not really so different. The key difference lies in how the kinetic energy operator T is "simulated." In GFMC, T is used in the propagator to cause diffusion, while in path-integral Monte Carlo its contribution becomes part of the potential—but between "replica" particles. Each approach has applications for which it is best suited. Path-integral Monte Carlo is particularly appealing aesthetically for understanding the Feynman sum-over-paths concept, but it is generally limited to low-dimensional quantum systems (frequently to only one electron).

We focus here on GFMC. Interest in GFMC as a computational tool arises from its ability to provide an exact stochastic solution to the many-body Schrodinger equation, particularly in a continuum. With importance sampling, highly accurate ground-[15–17] and excited-state energies,[18] as well as, more recently, other properties[19] of various atomic, molecular, and condensed quantum systems have been readily obtained. In fact, GFMC simulations are now being found frequently to be more accurate than *ab initio* quantum chemical calculations and local density methods. The ability to accurately treat quantum many-body correlations makes the approach very powerful, particularly because most other theoretical approaches have difficulty in this regard. Thus GFMC often provides both highly accurate answers and a check on other calculational procedures. Among its serious limitations, however, and thus also the impetus for much active current research in the method itself, are: 1) lack of a stable, exact algorithm for fermions; 2) large cpu requirements (characteristic of Monte Carlo in general); and 3) a very large power-law scaling of cost with atomic number of the constituent atoms being simulated. We will discuss later some of the recent work addressing this last point. We begin, however, by providing a simple presentation of GFMC from the perspective of the short-time approximation.

FORMALISM AND GREEN'S FUNCTIONS

By transforming the time-dependent Schrodinger equation

$$i\hbar\frac{\partial\Psi}{\partial t} = -\frac{\hbar^2}{2m}\nabla^2\Psi + V\Psi \tag{1}$$

to "imaginary time," one obtains (on the right-hand side) the sum of a diffusion term (with diffusion constant $D \equiv \hbar^2/2m$) and a rate term. Separately, both these processes have exactly-known solutions, each of which may be described by a stochastic process (a random walk and random branching respectively). But how is the solution of the imaginary-time equation related to the Schrodinger equation? Consider the formal solution of Eq. 1 in an eigenvalue expansion, namely

$$\Psi(\mathbf{R}, t) = \sum_i \varphi_i(\mathbf{R}) e^{-iE_i t/\hbar} \quad . \tag{2}$$

In imaginary time this takes the form of a sum of exponentially decaying terms. The ground state dominates the sum, and may be stabilized by shifting the zero of energy to E_0.

With this in mind, let us rewrite Eq. 1, in imaginary time, with shifted energy, (and also from now on with $\hbar$ absorbed into the unit of time), as

$$\frac{\partial\Psi}{\partial t} = D\nabla^2\Psi - (V - E_0)\Psi = -H\Psi \quad , \tag{3}$$

with $H = T + (V\text{-}E_0)$. The formal, Green's function (operator) solution to Eq. 3, is

$$G = e^{-tH} \quad . \tag{4}$$

The propagator is obtained by writing G in a position representation, namely

$$G(\mathbf{R} \to \mathbf{R}', t) = \left\langle \mathbf{R}' \middle| e^{-tT - t(V - E_0)} \middle| \mathbf{R} \right\rangle \ . \tag{5}$$

It was one of Kalos' important contributions to show how one could use Monte Carlo to sample from the exact G, even when it is unknown.[4–7] The method may be described as a Monte Carlo iteration of the Dyson equation (which takes the form of a Fredholm integral equation), starting with an exact G_0 for a simple potential V_0. Monte Carlo is used to sum the infinite Dyson series by a random walk. However, rather than perform this Monte Carlo within a Monte Carlo, it is often adequate to take a short-time approximation to G, for which one can write an explicit functional form.[12,15] Starting with $G = e^{-tH}$, one may write the following expression accurate to $O(t)$, where the error results from V and T not commuting:

$$\begin{aligned} G(\mathbf{R} \rightarrow \mathbf{R}', t) &\cong \langle \mathbf{R}' | e^{-tV/2} e^{-tT} e^{-tV/2} | \mathbf{R} \rangle e^{tE_0} \\ &= e^{-t\{[V(\mathbf{R}')+V(\mathbf{R})]/2-E_0\}} \langle \mathbf{R}' | e^{-tT} | \mathbf{R} \rangle \\ &\equiv w(\mathbf{R},\mathbf{R}'; t) G_D(\mathbf{R} \rightarrow \mathbf{R}', t) \end{aligned} \qquad (6)$$

The propagator G_D is the solution to the ordinary diffusion equation, and its solution is exactly known, namely

$$G_D(\mathbf{R} \rightarrow \mathbf{R}', t) = (4\pi D t)^{-3N/2} e^{-(\mathbf{R}-\mathbf{R}')^2/4Dt} \qquad (7)$$

Thus, we see that Eq. 3 may be solved stochastically by the following procedure: Beginning with an ensemble of coordinates $\{\mathbf{R}\}$, which are 3N-dimensional points representing the positions of all the quantum particles being considered, move each $\mathbf{R}$ to $\mathbf{R}'$ with a Gaussian random number chosen from a distribution of width $2Dt$ and mean of zero. At $\mathbf{R}'$, weight the configuration by $w(\mathbf{R},\mathbf{R}';t)$. A convenient form of weighting, that in many instances increases sampling efficiency, is branching. One simply makes $\mathrm{int}(w + \xi)$ copies of the walker at $\mathbf{R}'$. Here ξ is a uniform random variate in the interval (0,1). One iteration of the ensemble then amounts to performing the following integration

$$\Psi(\mathbf{R}', t) = \int G(\mathbf{R} \rightarrow \mathbf{R}', t) \Psi(\mathbf{R}, t = 0)\, d\mathbf{R} \; . \qquad (8)$$

Regardless of the original distribution $\Psi(\mathbf{R},0)$, $\Psi(\mathbf{R},t)$ is now a solution to Eq. 3. However, from the eigenvalue expansion in Eq. 2, it is clear that one must iterate Eq. 8 to asymptotically large t to converge to the ground-state distribution.

This approach may be viewed as a continuum version of Ulam's original idea. As with the discrete version, this procedure is unfortunately still inefficient. Efficiency is generally improved when known information is incorporated into Monte Carlo, which we will do in a moment. However, another apparent problem with the above is the branching on the function $(V-E_0)$. Because the Coulomb potential is singular whenever a pair of particles approach one another, branching will sooner or later get out of control. Importance sampling is the solution to both these problems.

IMPORTANCE SAMPLING AND FERMI STATISTICS

For GFMC, Kalos found a simple, but remarkably effective importance sampling scheme,[8] obtained by multiplying the imaginary, time-dependent Schrodinger equation (Eq. 3) by a trial wave function, Ψ_T. Expressing this new equation in terms of the product $f = \Psi_T \Psi$, one is able to obtain

$$\frac{\partial f}{\partial t} = D\,\nabla^2 f + [E_0 - E_L(\mathbf{R})]f - D\,\nabla\cdot[f\,\mathbf{F}(\mathbf{R})] \quad , \tag{9}$$

where the quantities $E_L = H\Psi_T/\Psi_T$ and $\mathbf{F} = 2\nabla \ln \Psi_T$ are respectively the local energy and a drift velocity imposed on the diffusion. This equation retains many of the important features of Eq.3, in particular the resemblance to a diffusion plus branching process. However, in many respects, it is better behaved. The singularities in the branching factor are now virtually suppressed, since E_L (unlike V) is almost a constant. In fact, E_L approaches a constant as $\Psi_T \rightarrow \Psi$. Furthermore, the drift term now guides the walk preferentially into regions where the wave function is large.

To see the effect this has on the Green's function, consider the propagation of $\Psi(t)$ to $\Psi(t+\tau)$. Multiply Eq.8 (for initial time t, and an increment τ) by $\Psi_T(\mathbf{R}')$ to obtain

$$\begin{aligned} f(\mathbf{R}', t+\tau) &= \int \left[\Psi_T(\mathbf{R}')\, G(\mathbf{R} \rightarrow \mathbf{R}', \tau)\,\Psi_T^{-1}(\mathbf{R})\right] \Psi_T(\mathbf{R})\Psi(\mathbf{R}, t)\, d\mathbf{R} \\ &= \int \tilde{G}(\mathbf{R} \rightarrow \mathbf{R}', \tau)\, f(\mathbf{R}, t)\, d\mathbf{R} \quad . \end{aligned} \tag{10}$$

The Green's function $\tilde{G}$ propagates f, and is given by the bracketed expression above.

A more physically transparent form of $\tilde{G}$ may be derived as follows. If we write Eq. 9 for the Green's function as

$$-\frac{d\tilde{G}}{dt} = \tilde{H}\,\tilde{G} = (\tilde{T} + \tilde{V})\tilde{G} \quad , \tag{11}$$

with

$$\tilde{T} = -D\,\nabla^2 + D\,\nabla\cdot\mathbf{F} + D\,\mathbf{F}\cdot\nabla \tag{12a}$$

and

$$\tilde{V} = E_L - E_0 \; , \tag{12b}$$

we can reproduce the steps followed in Eqs. 4-7, to obtain

$$\tilde{G}(\mathbf{R} \rightarrow \mathbf{R}', t) = \omega(\mathbf{R}, \mathbf{R}'; t)\, \tilde{G}_D(\mathbf{R} \rightarrow \mathbf{R}', t) \quad . \tag{13}$$

The weight is now given by

$$\omega(\mathbf{R}, \mathbf{R}'; t) = \exp(-t\{[E_L(\mathbf{R}) + E_L(\mathbf{R}')]/2 - E_0\}) \quad , \tag{14}$$

and the diffusion with drift propagator

$$\tilde{G}_D = \langle \mathbf{R}' | \exp(-t\tilde{T}) | \mathbf{R} \rangle = (4\pi D t)^{-3N/2} e^{(\mathbf{R}-\mathbf{R}'-Dt\,\mathbf{F}(\mathbf{R}))^2/4Dt} \tag{15}$$

is the solution to Eq. 9 without the branching term. Thus, the importance-sampled procedure is essentially the same as the first scheme presented above, except the diffusion now has a superimposed drift, and the branching is modified. The simulational procedure, however, remains exactly the same. With or without importance sampling, a nagging problem arises when one considers quantum statistics. As we discussed briefly earlier, for fermions, such as electrons, the wave function must be antisymmetric. Therefore, when more than one electron is in the same spin state (up or down), the wave function must somewhere change sign. This is troublesome if one wants to interpret Eqs. 3 and 9 as diffusion equations, in which the density of diffusers, or walkers, must be positive. How to handle this situation is not fully solved to this day, though a number of workable approaches have been developed.[11,20,21] A generally very good approximation for treating Eq. 9 follows from noting that the distribution f will always remain positive if Ψ and Ψ_T change signs together. This is readily accomplished by forcing the solution Ψ to vanish at the nodes of Ψ_T. This "fixed-node approximation" is variational—i.e. the energy thus derived will be an upper bound to the true energy. Many other methods have been devised to treat fermions, though all appear to contain some drawbacks. The fixed-node approach has the virtue of being simple, stable, variational, and generally accurate. Use of several trial functions, or optimizing parameters of a single Ψ_T can give an indication of the order of magnitude of the approximation being made. The reader interested in pursuing the problem of Fermi statistics in GFMC is refereed to Refs. 11, 15 and 20.

CRITICAL SLOWING DOWN IN ATOMS

From a practical point of view, one of the most serious problems with GFMC simulations has been the difficulty associated with large atomic number constituents. It has been estimated[22–23] that the cpu cost rises roughly as the 6*th* power of Z (where the nuclear charge Z equals the atomic number for neutrals). This means that a simulation of atomic Ne ($Z = 10$) costs a million times as much as one for hydrogen, for similar statistical errors. Thus Ne has in a practical sense been a limit to what can be simulated. Let us take a moment here to examine the reasons behind this large scaling power.

Questions such as this are often examined from the point of view of computational complexity. For example, moving N electrons takes order N operations. At each point (at least in our algorithm) a matrix inversion must be performed. We use a pretty unsophisticated matrix inverter, which requires N^2 operations for an N by N matrix. Thus, order N^3 operations are taken. As $N \sim Z$, three of the 6 powers of Z are explained in this way. The remaining three powers of Z result from the scaling of time step size, τ, with nuclear charge. Intuitively, the inner orbits contract, while the outermost ones expand. Thus, in atomic systems, as in critical phenomena, there exists a hierarchy of scales, in this case deriving from the shell structure. Carrying this analogy further, as the nuclear charge $Z \to \infty$ we have a situation akin to a critical point: there ceases to be a largest energy scale, the level spacing spreads from zero to infinity, and there is a diverging length scale. Most importantly, there is a dynamic critical exponent, giving the divergence as a power of Z of the time scale (e.g. for an electron to make a Bohr "orbit"). It is here that the remaining three powers of Z will be found.

To make this scaling argument, consider the Bohr picture of a hydrogenic atom with charge Z. The radius of the orbit is given by

$$r = \frac{\hbar^2}{me^2}\frac{n^2}{Z} \quad , \tag{16}$$

and the velocity of the electron in the orbit is

$$v = \frac{e^2}{\hbar}\frac{Z}{n} \quad . \tag{17}$$

Combining these expressions, one obtains that the time to complete a Bohr orbit scales as

$$t \sim n^3 Z^{-2} \quad . \tag{18}$$

Thus the innermost electron (which sees a charge of Z, and has n=1) moves Z^2 faster than the electron in H. On the other hand, the outermost electron (which sees a charge of unity, but has quantum number n) moves n^3 slower than the electron in H. From consideration of how electrons fill the atomic shell structure, it is easy to see that $Z \sim n_{max}^3$, implying that the outermost electron moves Z times slower than the electron in H. Therefore, one needs to take steps Z^2 smaller to sample in the core region, but at the same time one needs to take Z more steps to sample the valence region. This accounts for the factor of Z^3. In effect, standard GFMC algorithms have a dynamic critical exponent $\nu z = 3$, where the exponents have their usual meanings.

An unusual feature here is the simultaneous expansion of the largest distance scale (by $Z^{2/3}$) with its corresponding slowing down (by Z), *and* a contraction of the smallest distance scale (by Z) with a *speeding up* of those modes (by Z^2). Hence different dynamic exponents can be defined by the relation

$$t_m = r_m^{-z_m}\,, \tag{19}$$

depending on the mode m. For the outermost electrons $z = 3/2$, while for the innermost $z = 2$. However, it is only the ratio of the largest to the smallest scales that is significant, at least for us here. As discussed above, that ratio for the times goes as Z^3. It is easy to confirm that the ratio of the lengths goes as $Z^{5/3}$, giving a dynamic critical exponent of $z = 9/5$.

ALGORITHMS FOR LARGE-Z

Pseudopotentials

A number of ideas have been explored for overcoming the large-Z problem, though with differing degrees of success. The most obvious is to replace the core electrons by a pseudopotential, leaving only the valence electrons to be simulated. This makes good physical sense, as predominantly the valence electrons participate in binding, dissociation, structure determination, reactivity, etc. Both the *ab initio* effective-core potential[23-24] and the model potential[25]—two well-known examples—have been applied to GFMC simulations. These are approximate, but both work very well for one and two valence electron systems. When additional valence electrons are added, singularities appear at the nodes causing simulational instabilities. A more promising pseudopotential-type approach has been specifically designed for GFMC. It is more properly called a pseudo-Hamiltonian method, because the kinetic energy operator is also modified.[26] Very promising results have been forthcoming with it, together with better simulational behavior, and the additional property of maintaining the variational nature of the energy.

Damped-core Monte Carlo

Besides methods which eliminate core electrons, some recent all-electron approaches also appear to show exceptional promise in speeding up large-Z simulations. One of these[27] treats the core by a fast and accurate (but nevertheless approximate)

Metropolis algorithm, while treating the important valence electrons with the full GFMC apparatus. This method does away with some of the limitations of pseudopotential methods, such as the neglect of polarization and correlation effects induced by interaction of the core and valence electrons. Nevertheless, the approach remains approximate because the core distribution is sampled from a trial distribution $|\Psi_{core}|^2$, and is not exactly generated. One may, nevertheless choose Ψ_{core} as accurately as desired, even including explicit interelectronic distance terms—a key ingredient for accurate wave functions. The (more important) valence electrons are sampled exactly with GFMC using a separate trial function Ψ_{val}. The valence Monte Carlo walk moves in the potential created by the core electrons. By damping the branching as walkers enter the core, a smooth crossover is effected from the exact GFMC valence walk to the Metropolis walk at the core. This damping gives the method its name.

In this scheme the valence electrons are unrestricted by the much faster core time scale, and so may take significantly larger steps τ than usually possible. The core evolves separately, enabling one to sample a representative distribution of cores. Thus the large Z dependence is reduced, both because the valence energy has a much weaker dependence on Z than the total energy, and because time steps appropriate for the valence electrons are readily used. Damped-core calculations have been done thus far for electron affinities and ionization potentials of a number of atoms, and dissociation energies of a few molecules. Calculated quantities show pleasing agreement with experiment. The largest Z system treated to date is Ge (Z=32) —significantly larger than any other system treated by an all-electron GFMC approach. In fact, this simulation should take 10^3 times longer than the previously mentioned limit of Ne. The reason this simulation is possible is that the damped-core Z-dependent speed-up for Ge gives a factor of 5000 (and more as Z goes up) over standard GFMC!

Acceleration

The similarity of the large-Z problem with critical slowing down, brings us back from quantum mechanics to the realm of statistical mechanics. So it makes sense to apply statistical mechanical techniques here. Recent approaches for overcoming critical slowing down in spin systems (as well as in quantum chromodynamics on a lattice) involve the use of acceleration algorithms.[28–29] To see how this might apply to GFMC, consider Eq. 9 with the branching suppressed.[30] One obtains then $\tilde{G} = \tilde{G}_D$ (cf. Eqs. 13, 15), with no weight. This Monte Carlo procedure samples the distribution $f = |\Psi_T|^2$. Sampling this distribution is also readily accomplished with a standard Metropolis algorithm, and is referred to as variational Monte Carlo (VMC) since it was originally used as a method of finding upper bounds to ground state energies and for optimizing trial wave functions. For VMC, the transformed imaginary-time Schrodinger equation is a Fokker-Planck (FP) equation for the probability distribution $f(\mathbf{R},t)$. Every FP equation has a corresponding Langevin equation describing the evolution of the dynamic variables of the distribution. This latter is a stochastic differential equation. In our case here, the time-discretized Langevin equation,

$$\mathbf{R}' = \mathbf{R} + D\tau\mathbf{F}(\mathbf{R}) + \chi\,, \tag{20}$$

is used to update the configurations $\{\mathbf{R}\}$ in the Monte Carlo procedure, and embodies the transition $\tilde{G}_D(\mathbf{R} \to \mathbf{R}', t)$. Here $\{\chi\}$ is a set of Gaussian random numbers (one for each component of $\{\mathbf{R}\}$) whose means are zero, and widths are $2D\tau$. The time-step bias introduced by discretizing the Langevin equation is removed by a Metropolis acceptance/rejection step.

It has been shown in a field-theoretic context[28] that one can modify the dynamics of the FP equation and still obtain the *same* steady-state solution. This is the basis of Fourier acceleration. Applied to VMC one obtains the FP equation [30]

$$\frac{\partial f}{\partial t} = DM_{ij}\frac{\partial}{\partial r_i}\frac{\partial}{\partial r_j}f - D\frac{\partial}{\partial r_i}M_{ij}F_j f\,, \tag{21}$$

and its associated Langevin equation

$$r'_i = r_i + D\tau M_{ij}F_j(\mathbf{R}) + \left(M^{1/2}\right)_{ij}\chi_j\,. \tag{22}$$

In these equations, repeated indices are summed over. Though the dynamics has been modified, the stationary state is the same regardless of the choice of the matrix $\mathbf{M}$, as long as $\mathbf{M}$ is real-symmetric, positive-definite, and independent of the coordinates $\{r_i\}$. Here r_i is the *ith* component of the 3N dimensional vector $\mathbf{R}$ (for N electrons). In other words, there are many different stochastic processes which relax to the same equilibrium, though they all do so at different rates. The key is that by appropriately choosing $\mathbf{M}$, one may "mix" modes so that normal coordinates all propagate at comparable speeds.

But how does one choose **M**? Since the equilibrium is unaffected by the choice, one is free to experiment. In fact, a simple choice turns out to work quite well. In the Langevin equation, **F** acts as a force. From the form of **F** (see below Eq. 9) it is derivable from a potential $U = -2\ln \Psi_T$. The configurations {**R**} seek the minima in this potential, but are not able to stay precisely at these minima because of the stochastic noise term in the Langevin equation. Assuming that near these minima the potential is harmonic, i.e. $U = a_{ij}q_i q_j/2$ in the normal coordinates {**q**}, one can see that the fast and slow motions in this potential correspond to large and small eigenvalues of a_{ij}. By choosing **M** as the inverse of the Hessian of U at a minimum (the Hessian is the matrix of second derivatives of U), the a_{ij} in **M** and **F** cancel in Eq. 22, resulting in the presence of only a single time scale—i.e. all modes relax at the same rate.

In simulations as small as Z=7 (N atom), the moves for individual electronic coordinates are already appreciably different with and without acceleration. For example, at typical time steps without acceleration, core steps are rejected almost without exception, while valence steps, conversely, are all accepted. The cause for this behavior is that a single time scale cannot cover both the valence and core regions effectively. In addition, as we saw earlier, these regions diverge from each other with increasing Z. Though a time step may be large for the core, making almost any move go into a statistically less favorable region (leading to rejection), in the valence region this same step is too small, and the electrons do not move appreciably in the available space. Thus the difficulty of large Z simulations may be seen to result from the impossibility of balancing these growing opposing constraints. The "ideal" acceptance ratio results in unacceptably slow convergence, as it takes exponentially long for the core electrons to relax. When acceleration is turned on we find that large and comparably sized steps are taken for core and valence electrons. In addition, after approximately 10-20 time steps, the core and valence electrons interchange. Acceleration factors of from 30-50 have been observed already at Z=7. The actual figure may be much greater still since, as indicated above, true equilibrium without acceleration could take exponentially long with the time steps typically used, or would require a much smaller τ before the core electrons relaxed.

We have only discussed accelerating VMC. For the full GFMC one must reintroduce the branching factor. This is not as trivial as it might seem, as branching too depends on the time step, and hence on the dynamics. What time scale should go into the branching factor $w(\mathbf{R},\mathbf{R}';t)$? Although a mathematically justifiable procedure for accelerating GFMC with branching has not yet been achieved, acceleration ideas have been used to decorrelate moves[30] without actually changing the dynamics of Eq. 9. Consider again the Green's function of Eqs. 13-15, and let us rewrite it, inserting a complete set of states, as

$$\langle \mathbf{R}'|\tilde{G}|\mathbf{R}\rangle = \langle \mathbf{R}'|\exp(-\tau \mathcal{T}_A)|\mathbf{R}''\rangle\langle \mathbf{R}''|\exp\{-\tau(\mathcal{T}-\mathcal{T}_A)\}|\mathbf{R}\rangle\, \tilde{w}(\mathbf{R},\mathbf{R}';t)\ . \tag{23}$$

The accelerated kinetic energy operator is just the same as used above for VMC, namely

$$\mathcal{T}_A = DM_{ij}\left[-\frac{\partial}{\partial r_i}\frac{\partial}{\partial r_j} + \frac{\partial}{\partial r_i}F_j + F_j\frac{\partial}{\partial r_i}\right]. \tag{24}$$

The full propagator is now the iteration of two propagators followed by a weight. Algorithmically, this means that the ensemble of configurations {**R**} is first updated to the intermediate ensemble {**R''**} by a matrix of (**1** - **M**), from which the accelerated Green's function then propagates the ensemble to {**R'**}. Although the combination of the two propagators formally undoes the acceleration, the decorrelation of the moves remains.

PERSPECTIVE

Recent years have witnessed a fairly successful first round of attack on the large-Z problem. The use of hybrid approaches (e.g. Metropolis together with GFMC or Langevin-based acceleration and GFMC) has been particularly powerful. One can look forward to additional ideas for speeding up GFMC, as well as applications to atomic, molecular, and condensed-matter systems in which atoms of carbon and beyond are routinely simulated. This would ultimately open the door to applications in areas ranging from clusters to materials science. Acceleration in particular appears to offer the hope for an exact treatment of electronic structure without the need for inordinate computer resources. The connections between large-Z and critical behavior serve both as an aesthetically pleasing bridge back to statistical mechanics, and as a useful tool for improving GFMC.

ACKNOWLEDGEMENTS

It is a pleasure to acknowledge fruitful collaborations and interactions with many people, especially Berni Alder, Rob Barnett, George Batrouni, David Ceperley, Brian

Hammond, and William Lester, Jr. Additional thanks go to my wife, Louise Perini, for a critical reading of this paper.

REFERENCES

1] See discussion in *Symposium on Monte Carlo Methods*, H. A. Meyer, ed. (Wiley Interscience, New York,1956).

2] N. Metropolis and S. M. Ulam, J. Am. Stat. Assoc. 44, 247 (1949), and N. Metropolis in Ref. 1].

3] N. Metropolis, A. W. Rosenbluth, M. N. Rosenbluth, A. H. Teller, and E. Teller, J. Chem. Phys. 21, 1087 (1953).

4] M. H. Kalos, Phys. Rev. 128, 1891 (1962); M. H. Kalos, Nucl. Phys. A 126, 609 (1969).

5] M. H. Kalos, J. Comp. Phys. I, 257 (1966); M. H. Kalos, Phys. Rev. A 2, 250 (1970).

6] G. E. Albert, in Ref. 1].

7] K. S. Liu, M. H. Kalos, and G. V. Chester, Phys. Rev. A 10, 303 (1974); P. A. Whitlock, D. M. Ceperley, G. V. Chester, M. H. Kalos, Phys. Rev. B 19, 5598 (1979); D. M. Ceperley and M. H. Kalos in *Monte Carlo Methods in Statistical Physics*, K. Binder, ed. (Springer-Verlag, Berlin, 1979).

8] M. H. Kalos, D. Levesque, and L. Verlet, Phys. Rev. A 9, 2178 (1974).

9] M. H. Kalos, M. A. Lee, P. A. Whitlock, G. V. Chester Phys. Rev. B 24, 115 (1981); K. E. Schmidt, M. H. Kalos, M. A. Lee Phys. Rev. Lett. 45, 573 (1980); P.A. Whitlock, M. H. Kalos, J. Comp. Phys. 30, 361 (1979).

10] D. M. Ceperley, J. Comp. Phys. 51, 404 (1983).

11] D. M. Ceperley and B. J. Alder, Phys. Rev. Lett. 45, 566 (1980); D. M. Ceperley and B. J. Alder, J. Chem. Phys. 81, 5833 (1984).

12] D. M. Ceperley in *Recent Progress in Many-Body Theories*, J. G. Zabolitzky, M. de Llano, M. Forbes, and J. W. Clark, eds. (Springer-Verlag, Berlin, 1981).

13] J. B. Anderson, J. Chem. Phys. 63, 1499 (1975); 65, 4121 (1976); 73, 3897 (1980); Int. J. Quant. Chem. 15, 109 (1979).

14] L. Schulman, *Techniques and Applications of Path Integration*, (Wiley, New York, 1981); D. Chandler, *Introduction to Modern Statistical Mechanics*, (Oxford University Press, New York, 1987).

15] P. J. Reynolds, D. M. Ceperley, B. J. Alder, and W. A. Lester, Jr., J. Chem. Phys. 77, 5593 (1982).

16] R. N. Barnett, P. J. Reynolds, and W. A. Lester, Jr., J. Chem. Phys. 84, 4992 (1986); D. R. Garmer and J. B. Anderson, *ibid* 86, 7237 (1987); F. Mentch and J. B. Anderson, *ibid* 80, 2675 (1984); D. R. Garmer and J. B. Anderson, *ibid* 86, 4025, 7237 (1987).

17] R. J. Harrison and N. C. Handy, Chem. Phys. Lett. 113, 257 (1985).

18] R. M. Grimes, B. L. Hammond, P. J. Reynolds, and W. A. Lester, Jr., J. Chem. Phys. 85, 4749 (1986); D. M. Ceperley and B. Bernu, *ibid* 89, 6316 (1988).

19] B. H. Wells, Chem. Phys. Lett. 115, 89 (1985); P. J. Reynolds, R. N. Barnett, B. L. Hammond, and W. A. Lester, Jr., J. Stat. Phys. 43, 1017 (1986); A. L. L. East, S. M. Rothstein, and J. Vrbik, J. Chem. Phys. 89, 4880 (1988); V. Dobrosavljevic, C. W. Henebry, and R. M. Stratt, J. Chem. Phys. 91, 2470 (1989); R. N. Barnett, P. J. Reynolds, and W. A. Lester, Jr., "Algorithms for Calculating Quantum Properties by QMC," and "Transition Dipole Moment Calculations by Quantum Monte Carlo Methods," *preprints* (for J. Comp. Phys.)

20] D. M. Arnow, M. H. Kalos, M. A. Lee, and K. E. Schmidt J. Chem. Phys. 77, 5562 (1982).

21] D. M. Ceperley and B. J. Alder, Science 231, 555 (1986).

22] D. M. Ceperley, J. Stat. Phys. 43, 815 (1986).

23] B. L. Hammond, P. J. Reynolds, and W. A. Lester, Jr., J. Chem. Phys. 87, 1130 (1987).

24] M. M. Hurley and P. A. Christiansen, J. Chem. Phys. 86, 1069 (1987); P. A. Christiansen, J. Chem. Phys. 88, 4867 (1988).

25] T. Yoshida and K. Iguchi, J. Chem. Phys. 88, 1032 (1988).

26] G. B. Bachelet, D. M. Ceperley, and M. G. B. Chiocchetti, Phys. Rev. Lett. 62, 2088 (1989).

27] B. L. Hammond, P. J. Reynolds, and W. A. Lester, Jr., Phys. Rev. Lett. 61, 2312 (1988).

28] G. G. Batrouni, G. R. Katz, A. S. Kronfeld, G. P. Lepage, B. Svetitsky, and K. G. Wilson, Phys. Rev. D 32, 2736 (1985); E. Dagotto and J. Kogut, Phys. Rev. Lett. 58, 299 (1987); G. G. Batrouni and B. Svetitsky, Phys. Rev. B 36, 5647 (1987); G. G. Batrouni and E. Dagotto, *ibid* 37, 9875 (1988).

29] R. Swendsen and J. S. Wang, Phys. Rev. Lett. 58, 86 (1987); U. Wolff, Phys. Rev. Lett. 60, 1461 (1988); W. Klein, T. Ray, and P. Tamayo, Phys. Rev. Lett. 62, 163 (1989).

30] G. G. Batrouni and P. J. Reynolds, "Accelerating Green's Function Monte Carlo: Avoiding Critical Slowing Down in Simulations Containing Large-Z Atoms," Phys. Rev. Lett. *submitted.*.

Modified Q2R Cellular Automata

Paulo Murilo C. de Oliveira, Thadeu J. P. Penna,
Rita M. Z. dos Santos and Suzana M. M. de Oliveira
Instituto de Física, Universidade Federal Fluminense
C.P. 100296 Niterói RJ, Brazil 24000

ABSTRACT

Two different versions of cellular automata are studied in the square lattice, both in connection with the microcanonical Ising model behavior. For the first version, we get improved values for both the magnetization and susceptibility, as compared with the Q2R counterparts. For the second version, we found a new transition for which the critical point and critical exponents are numerically determined.

The so–called Q2R cellular automaton was proposed[1] as a microcanonical deterministic simulator for the Ising model. A very fast computer implementation[2] was performed in two dimensions, and the measured magnetization seemed to confirm the initial proposal. Measures of the three dimensional susceptibility[3], however, were found far below the correct values in the ordered phase, although good estimates were found above the critical point. Also, values for other thermodynamical quantities are well reproduced below and above the critical point. We found the same behavior also in two dimensions.

A possible reason for the underestimated susceptibility in the ordered phase is the low degree of fluctuation allowed by the Q2R restricted rule (the susceptibility is a measure of the fluctuations of the magnetization). We recall that each spin is flipped, following the Q2R rule, if and only if just half of its nearest neighbors are parallel to it. The *local* energy, besides the total value for the whole lattice, is thus conserved in this dynamical process: the energy contained in antiparallel-spin-bonds can migrate only through one-lattice-parameter steps. The Q2R rule did not allow the annihilation of energy at some place, and the simultaneous creation of an equal amount of energy at another place of the same lattice. Following this reasoning, we defined some alternative cellular automaton rules, all of them similar to Q2R. We report now some results obtained from two of our rules in two dimensions (square lattice), in a very brief way. We measured both magnetization and susceptibility.

First rule:

Each pair of nearest neighbor spins is flipped (both spins at once) if and only if just half of its six surrounded bonds link parallel spins. We found improved results for both the magnetization and the susceptibility, compared to Q2R, but yet the ordered phase susceptibility is underestimated. We performed simulations on 128×128 lattices. Note that the energy-carrier bonds can now migrate over distances larger than one lattice parameter, but the local character of the energy exchange is still present.

Second rule:

After one complete Q2R update of the whole lattice (normal Q2R rule), we choose a 4-energy spin (one with 4 antiparallel neighbors) and another 0-energy spin (one with 4 parallel neighbors) — only one spin of each kind, if they exist — and both are flipped. The whole process (one complete Q2R update of the whole lattice and the subsequent flipping of one 4- and other 0-energy spins) is repeated iteratively. Note now the non-local character of the energy exchange: the search for the 4-energy spin is performed in a systematic sequence along the lattice, starting from the last one found in an earlier iteration; the same prescription for the 0-energy spin. The result was a surprise: the Curie transition disappears, and another transition at a smaller critical energy was found. From our numerical data, this critical value seems to be the same as the "cluster period" critical point of the transition from non-periodic to periodic behavior observed in Q2R[4]. We performed simulations on 64×64, 256×256 and 512×512 lattices. Through data–collapsing plots of susceptibility and magnetization, we found the same set of exponents of the ordinary Curie Ising transition.

References

1) Y. Pomeau, *J. Phys.* **A17**, L415 (1984).

2) H. J. Herrmann, *J. Stat. Phys.* **45**, 145 (1986);
J. G. Zabolitzky and H. J. Herrmann, *J. Comput. Phys.* **76**, 426 (1988).

3) W. M. Lang and D. Stauffer, *J. Phys.* **A20**, 5413 (1987).

4) H. J. Herrmann, H. O. Carmesin and D. Stauffer, *J. Phys.* **A20**, 4939 (1987).

Work partially supported by brazilian agencies CAPES, FINEP and CNPq.

A CELLULAR AUTOMATA FOR THE ORDER-DISORDER TRANSITION

Maria Clara L.Adams and José R.Iglesias
Instituto de Física
Universidade Federal do Rio Grande do Sul
Caixa Postal 15051, 91500 Porto Alegre, RS
Brasil

ABSTRACT

We try to develop a cellular automata that describes the dynamics of an Ising system conserving total magnetism. If the interactions are antiferromagnetic the order-disorder transition of a binary alloy is described. Two simple rules are presented and the results discussed. A third rule is proposed in order to strictly guarantee the conservation of the magnetization.

Cellular automata (1,2) are widely used to study the Ising model. Specifically the automata Q2R (3) has been proposed to simulate the thermodynamics of the Ising model (4,5). It is a deterministic automata that strictly conserves energy.

Here, we are interested in the order-disorder transition. As it is well known, some binary alloys at or near the 50-50 stoichiometry can have a transition from a disordered (high temperature) phase to an ordered (low temperature) phase. Then we look for a system where the number of A and B species (spin up = A and spin down = B) is constant. That's the reason why we tried a deterministic rule to simulate the dynamics of the order-disorder transition. The mechanism of the transition is a diffusion of atoms from one site to another and, as atoms are not created, neither destroyed, we should have a constant magnetization.

Two rules were tried. The main idea in both was to consider the A and B sites equivalent so that the rule was symmetric in relation to the up and down spins. The update of a site will be permited only if in the vicinity exists at least one of the same specie.

Rule 1 is one of the rules that considers only nearest neighbours (rule Q1Q0124 in the classification of Vichniac (ref. 3)). At each time step the spins of the sites are updated if the local energy is lowered. For each site the local energy is:

$$E_i = -J\, s_i \sum_j s_j$$

where s_i is the value of the spin (type of atom) and J the interaction energy (negative for antiferromagnetism).

The results are exhibit on fig. 1 starting from an arbitrarly disordered phase. Also the total energy as a function of the number of steps is plotted on fig. 2.

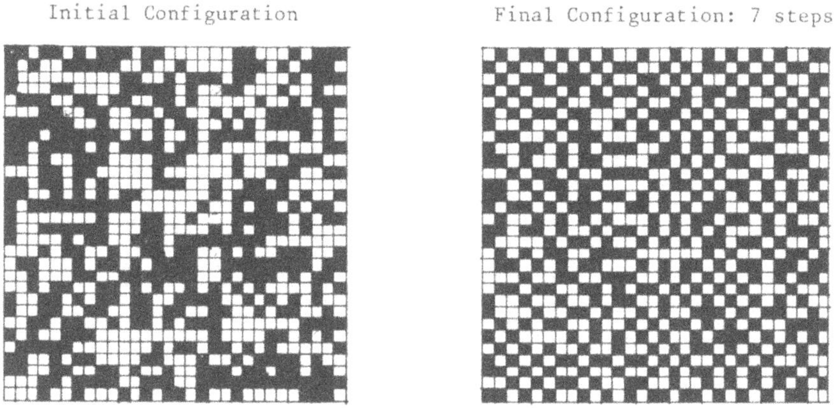

(a) (b)

FIGURE 1

This rule is unable to generate a full ordered phase and the fluctuation in the number of A and B atoms is too high.

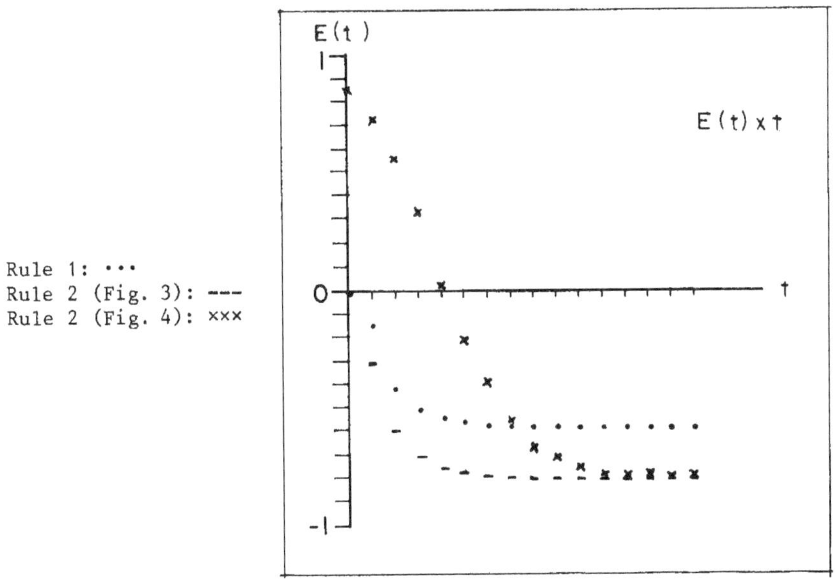

FIGURE 2: Energy as a function of the number of time steps.

We essaied a second rule of a new type: it takes into account also the second neighbours. The dynamics is performed in two steps: initially one considers the first four neighbours and, if no modification is induced by the first rule, one looks to the second neighbours. Here again we will only update the site if thislowers the energy. In both rules, if all the neighbours are of the same specie, we do not update the central spin (although it will decrease the energy) in order to not "create" new atoms.

On fig. 3 we present the results, starting from an disordered phase, and in fig. 4 starting from a region of segregation. The energies as a function of time are also plotted in fig. 2.

Although we are not able to conserve the magnetization in all steps, the maximum deviation is 4.22% and the final results show a 2.0% deviation from the initial. For the disordered initial configuration the results are better 2.89% and 0.44% respectively.

Rule 2 seems to describe fairly well the transition. A lower value of the energy is obtained and well-defined ordered regions are evident on figs. 3b and 4b, separated by grain-boundaries.

This automata can also be used to study diffusion and it is easy to generalize it for 3 dimensions.

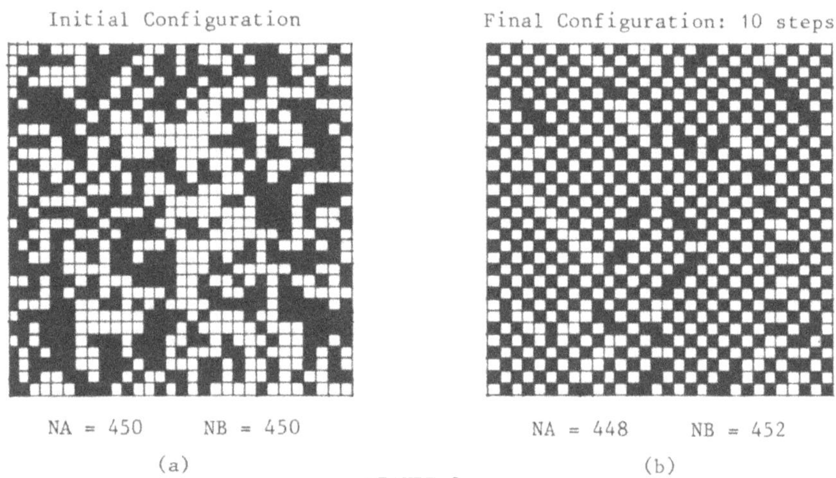

FIGURE 3

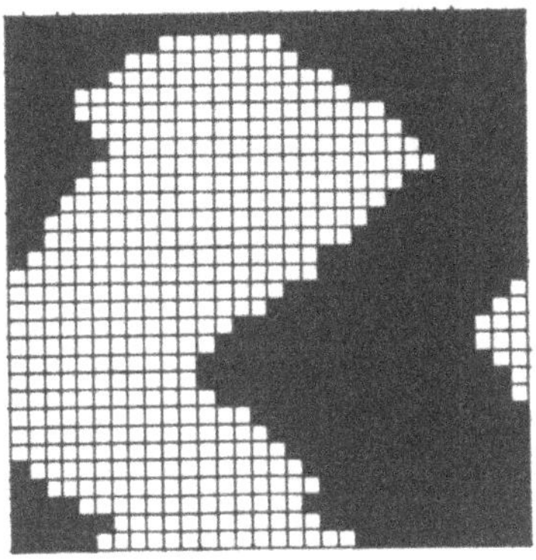

(a)

(b)

FIGURE 4

Finally, a more physical description of those systems can be obtained by a probabilistic automata that rigorously conserves the magnetization although it would cost more computer time. In this case the units to be considered consists, for example, of two neighbours sites randomly oriented. These units cover all the lattice. The spins of the sites should be exchanged if this lowers the energy of the plates in contact. A probability of spontaneous exchange can be considered through a Boltzmann factor.

ACKNOWLEDGEMENT

This work originates in a seminar by H.Herrmann. We acknowledge discussions with R. de Almeida and J.Arenzon and the financial support of brazilian agency CNPq.

REFERENCES

1. Wolfram, S., "Theory and Applications of Cellular Automata", World Scientific, Singapore (1986).
2. Rujan, P., J. Sat. Phys. 49, 137 (1987).
3. Pomeau, Y., J. Phys. A: Math. Gen. 17, L415 (1984); Vichniac, G.Y., Physica 10D, 96 (1984).
4. Herrmann, H.J., J. Stat. Phys. 45, 145 (1986).
5. Herrmann, H.J., Carmesin, H.O. and Stauffer, D., J. Phys. A: Math. Gen. 20, 4939 (1987).

KINETICS OF ION BEAM INDUCED AMORPHISATION

J.A.T. Borges da Costa
Departamento de Física - Universidade Federal de Santa Maria, 97119, Santa Maria, RS, Brasil

C. Scherer
Instituto de Física - Universidade Federal do Rio Grande do Sul, C.P. 15051, 91500, Porto Alegre, RS, Brasil

ABSTRACT

The problem of covering a limited space with randomly placed space-overlapping geometrical elements is here addressed as a model description of the mechanism by which the amorphisation of a region near the surface of semiconductors or metallic alloys is induced by heavy ion bombardment. Some of the results of a computer simulation are presented which are in close qualitative agreement with an experimental investigation of the kinetics of ion beam induced amorphisation.

INTRODUCTION

The problem of how many mutually exclusive cubes of volume v are required to completely cover a limited space of volume V (V>>v) has a trivial solution, namely, n=V/v. If instead of cubes one has, for instance, spheres, the answer to the same question is that it is impossible to cover the space if the spheres are not allowed to overlap. Now, allowing for overlapping, many solutions can be envisaged depending on further restrictions regarding the spatial arrangement of the geometrical elements. The problem of interest here is the one of random locations of those elements. The question is then: how many possibly overlapping elements of given shape and volume must be placed at random over a limited space in order to cover it completely?

A physical situation related to this problem is the low temperature amorphisation of a shallow portion near the surface of semiconductors [1] or metallic alloys [2] submitted to heavy ion bombardment. Due to the stochastic nature of the interactions between a beam of energetic ions and a solid system, the physical processes induced by each ion take place at random locations in the system. The time duration of those processes ($\cong 10^{-11}$ s) is many orders of magnitude smaller than the characteristic time separation between two successive disturbances of the medium at the same point. Complete amorphisation of the region near the surface of a solid is achieved only after the impingement of a number of ions per unit area, i.e., above a certain implanted dose. Therefore, it seems natural to understand low temperature ion induced amorphisation as a cumulative process resulting from the random coverage of the bombarded region by localised instantaneous transformations. In particular, experimental evidence indicates that, under certain conditions, amorphous clusters may be formed by each ion impact [3-4]. In the following two sections we present and analise the results of a computer simulation study of the kinetics of a transformation from the crystalline to the amorphous state by this later mechanism.

THE COMPUTER SIMULATION METHOD

A computer simulation code was designed which simulates the random covering of the space by geometrical elements. The simulations were carried out over a matrix of 50x50x100 cells in which the covering elements, choosen to be paralelepipeds for simplicity, were placed at random. The statistical location of each element was taken as independent of the others, i.e., overlapping was allowed. As we shall see later, the covering kinetics does not depend on the geometry of the covering elements but only on their volumes. The initial state of the matrix was set to be "0" (indicating that the system was crystalline). The state of the cells covered by an element was simply turned into "1" (indicating that the system became locally amorphous). After n steps, the sum of the contents of the cells, divided by the total number of cells, indicated the volume fraction, c, of the matrix covered at least once (the volume fraction of the amorphous phase).

The boundary conditions were choosen in such a way as to keep close proximity with typical ion bombardment experimental conditions. During such an experiment the range of the ions is limited in the direction of incidence of the ion beam by the energy loss during electronic and atomic collisions within the solid [5] (typical mean ranges are of the order of some hundreds of Angstroms). Therefore , the direction of the simulation matrix containing 100 cells was identified with the direction of the beam and the boundaries were made closed, i.e., no covering element could extend to outside that range.However, the lateral dimensions of the bombarded region are much greater (an area of about $1cm^2$ is commonly spanned by by the ion beam) and, in fact, can be considered as infinite. The lateral boundaries of the simulation matrix were then made periodic, i.e., the portion of a covering element extending to ouside a lateral boundary was continued in the oposite side of the matrix as if another identical covering element were placed in an equivalent location of a lateral periodic extension of the matrix.

The probability distribution of the covering elements

was assumed to be uniform over the entire matrix. Considering the ion induced amorphisation process, this assumption is good when applied to the lateral directions of the matrix and is complementary to the assumption of periodic boundaries in order to reproduce the condition of lateral uniformity of the ion bombardment. However, when applied to the direction of the ion beam, it will require some care in the analysis of the results. Indeed, a depth dependent probability distribution is expected for the location of the ion induced amorphous clusters. Therefore, the computer simulation results presented bellow have to be considered under this restriction which, together with the problem of the transmission of a fraction of the incident ions through the solid system, does not allow a quantitative comparison with experimental results even though good qualitative agreement is obtained. Nevertheless, the approximation is valid when the probability distribution is flat or when a limited region under a solow varying distribution is considered.

SOME COMPUTER SIMULATION RESULTS COMPARED WITH AN EXPERIMENTAL STUDY

Figure 1.a) shows the volume fraction, c, of the simulation matrix covered after the random location of n geometrical elements of two different sizes (7x7x7 and 13x13x13 cells) in a Log c versus Log n plot.

Complete coverage of the matrix requires about $10^{3.5}$ elements of size 7x7x7 which is nearly ten times the corresponding required number (about $10^{2.5}$) of 13x13x13 elements. In the low n portion of of the plot two straight lines are obtained with the same slope. There, the value of c in the upper curve is about $10^{-0.8}$ $\approx$6.3 times greater than the value obtained from the lower curve for the same n. This is approximately the ratio of the corresponding volumes of the covering elements. In fact, the two curves colapse into one single curve if one chooses n*v, instead of n, as the independent variable. The same was observed for all choices of the geometry of the covering elements. Thus, as we have already pointed out, one concludes that the covering kinetics does not depend on the geometry of the elements. The value of c is completely determined by their volumes.

In a paper by C. Jaouen, J. Delafond and J.P. Rivière [2], the transformation from the crystalline to the amorphous state of NiAl thin films has been studied by in situ electrical resistivity measurements during ion irradiation at 77K. In particular the amorphisation kinetics induced by 540 keV Bi^{+++} and 120 keV Ar^{+} irradiation is followed as a function of the implanted dose. The results are shown in figure 1.b) in a plot of log c versus log ϕ, where c is the amorphous fraction and ϕ is the implanted dose.

The authors propose that amorphisation is induced by a

localised process. The volume of the portion of the solid amorphized by each ion impact is estimated to be about 719nm for Bi and 111 nm for Ar. Hence the volume of the amorphous cluster generated by a Bi ion impact is in the same proportion to the volume of the amorphous cluster generated by an Ar ion impact as the ratio between the volumes of the covering elements of our computer simulation.As one can readily see, all the other kinetic parameters of the two ion bombardment experiments are in the same poportion as the characteristic parameters of our simulations. These are: the ratio between the threshold doses, ϕ, for complete amorphisation by Bi irradiation and by Ar irradiation is nearly 10 ((10^{15} ions/cm^2 (Bi))/(10^{14} ions/cm^2 (Ar))); the amorphous fraction , c, corresponding to the same ϕ in the low dose region is about 6.3 times greater for Bi irradiation than for Ar irradiation.

Figure 2 shows Avrami's plots [6] of the data from figure 1. A well defined kinetic law which conforms an equation of the form $c(t)=1-\exp(-at^{\nu})$, where t=n for the simulations and $t=\phi$ for the experiments, is apparent from the plots. In both cases, and under varied simulation or experimental conditions the exponent, ν , was found to be $\cong$1. This behaviour is characteristic of an homogeneous transformation kinetics controlled by a single mechanism.

CONCLUSIONS

In conclusion, the covering of a limited space by randomly localised geometrical elements which are allowed to overlap reproduces qualitatively the kinetic behaviour of low temperature amorphization of the NiAl thin film induced by heavy ion bombardment . The computer simulation code can be easily extended to investigate other influences on the amorphisation kinetics as, for example, the in-depth dependece of the amorphous clusters distribution or the need of disturbing (covering) a given region of the solid many times in order to induce amorphisation by some mechanism of defect acumulation [7-8].

ACKNOWLEDGEMENTS

The authors express their gratitude to M.A.Z. Vasconcellos for many suggestions and helpfull discussions, I.J.R. Baumvol and W.H. Schreiner for encouragement and material support in the Laboratorio de Filmes Finos - Instituto de Física - UFRGS, and S.R. Teixeira for his help in the preparation of the figures.

REFERENCES

[1] - Yarkulov, U., Rad. Eff. 100, 11(1986); 103, 135(1987)
[2] - Jaouen, C., Delafond, J. and Rivière, J.P., J. Phys.F: Met. Phys. 17, 335(1987)
[3] - Howe, L.M. and Rainville, M.H., Nucl. Instr. and Meth. 182/183, 143 (1981)
[4] - Ruault, M.O., Chaumont, J., Penisson, J.M. and Bouret, Phil. Mag. A50, 667(1980)
[5] - Ziegler, J.F., Biersack, J.P. and Littmark, U., "The Stopping and Range of Ions in Solids", Pergamon Press, Oxford, 1985.
[6] - Christian, J.W., "The Theory of Transformations in Metals and Alloys", 2nd. Ed., Pergamon Press, Oxford, 1975
[7] - Gibbons, J.F., Proc. IEEE 60, 1062(1972)
[8] - Drígo, A.V., Berté, M., Benyagoub, A., Bernas, H., Pivin, J.C., Pons, F., Thomé, L. and Cohen, C., Nucl. Instr. and Meth. 19/20, 533(1987)

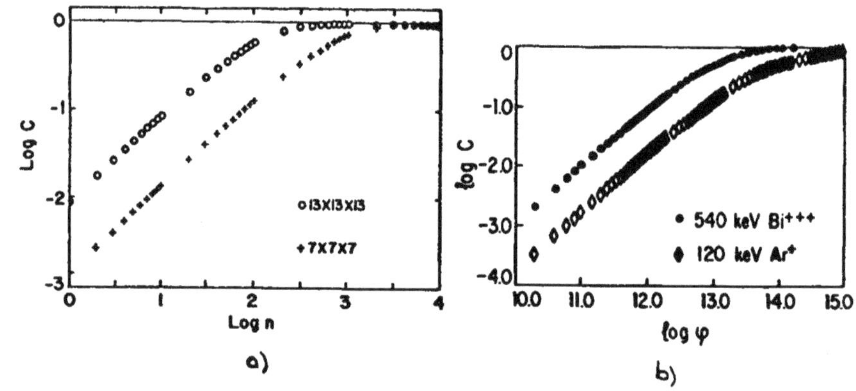

Fig. 1 - a) Volume fraction, c, of a matrix of 50x50x100 cells covered after the random location (with uniform probability distribution) of n statistically independent geometrical elements of sizes 13x13x13 O and 7x7x7+; b) volume fraction of a NiAl film (72nm thick) amorphised after the incidence of Ø energetic particles per cm^2: 540keV Bi^{+++} ●, and 120 keV Ar^+◊.

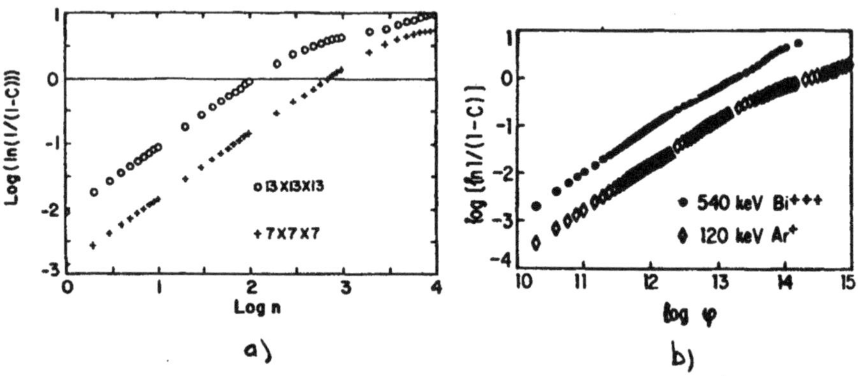

Fig. 2 - a) and b) are Avrami's plots of the data from figures 1.a) and 1.b) respectivelly.

SPACE OF PARTITIONS AND COMPLEXITY

Mario Casartelli
Dipartimento di Fisica dell'Università
and Unità CNR-CISM - Parma (Italy)

Abstract: We introduce a mathematical scheme, based on the metric and combinatorial properties of a space $\mathcal{Z}$ of measurable partitions, in order to analyze the complexity of Cellular Automata. This approach requires the embedding of $\mathcal{Z}$ in a space $\mathcal{R}$ of "rational partitions", where an extended entropy functional may be defined. Using such an entropy, it is possible to build up complexity indicators which, at least in one dimensional cases, can be numerically tested. Adaptations of these indicators to general discrete dynamical systems is almost immediate.

Cellular Automata (CA) offer a relatively simple field where general concepts arising in the study of complexity may be formulated and tested[1]. We introduce here an approach to the problem of a quantitative estimate of the complexity of CA, based on the metric and combinatorial properties of the space $\mathcal{Z}$ of finite partitions. This will be done in two steps: first, a short account about the properties of $\mathcal{Z}$ and the way to situate the complexity problem in it; second, the connection with CA.

As it is well known[2], the set $\mathcal{Z}$ of finite partitions $\alpha, \beta, \gamma ...$ of a set $\mathbf{M}$ constitute a metric space, where the distance $\rho(\alpha, \beta)$ is given by

$$\rho(\alpha, \beta) = H(\alpha|\beta) + H(\beta|\alpha).$$

$H(\alpha|\beta)$, the conditional entropy of α with respect to β, is a measure of the residual incertitude about α when β is known, and therefore ρ is a measure of the similarity of its arguments, in the sense that $\rho(\alpha, \beta)$ very small $\Rightarrow$ α and β very similar.

If $\alpha = \alpha'\sigma, \beta = \beta'\sigma$ (using for brevity $\alpha'\sigma \equiv \alpha' \vee \sigma$), it is possible that $\rho(\alpha, \beta)$ is small and $\rho(\alpha', \beta')$ is large: this means that the "common factor" σ is such to mask the difference between α' and β'. The problem of the reduction of the common factor σ is made difficult, with respect to the analogous arithmetical problem, by the fact that a uniquely defined factorization of a partition into prime factors does not exist. We shall give here a brief account of a reduction procedure which will be described elsewhere in detail.

Let $\alpha, \beta \in \mathcal{Z}$, and $\sigma = \alpha \wedge \beta$. Then the collections $\{\alpha_k\}$ and $\{\beta_j\}$ of partitions

$$\alpha_k \wedge \sigma = \beta_j \wedge \sigma = \nu; \quad \alpha_k < \alpha, \quad \beta_j < \beta$$

(ν being the unit partition) are well defined. Introducing then

$$\alpha' = \vee_k \alpha_k, \quad \beta' = \vee_j \beta_j,$$

π will denote the following "projection" operating on couples of partitions:

$$\pi(\alpha, \beta) = (\alpha', \beta').$$

Clearly, $\alpha' < \alpha, \beta' < \beta$; moreover it may be simply proven that $\pi(\alpha', \beta') = (\alpha', \beta')$ i.e. $\pi^2 = \pi$. Such a reduction from (α, β) to (α', β') is uniquely defined. By analogy with operations on natural numbers defining the rational field, we call "rational partitions", or r-partitions, the couples in the range $\mathcal{R}$ of π. Note however that it may happen that $\alpha' \wedge \beta' \neq \nu$. If $\mathbf{a} = (\alpha_1, \alpha_2), \mathbf{b} = (\beta_1, \beta_2)$ are in $\mathcal{R}$, then

$$\mathbf{a} \vee \mathbf{b} \equiv \mathbf{ab} = \pi(\alpha_1\beta_1, \alpha_2\beta_2)$$

This extends the product "$\vee$" from $\mathcal{Z}$ to $\mathcal{R}$. Precisely, $\mathcal{Z}$ results to be isomorphous to the set $\mathcal{R}^\nu \subset \mathcal{R}$ of the couples (α, ν). If two r-partitions $\mathbf{a} = (\alpha_1, \alpha_2), \mathbf{b} = (\beta_1, \beta_2)$ are such that $\mathbf{a} \vee \mathbf{b} = (\alpha_1\beta_1, \alpha_2\beta_2)$ (i.e. if the reduction has no effects) then they are called "relatively independent", or r.i. r-partitions. $\mathcal{R}$ is endowed with very peculiar features, e.g. the extended product, which in $\mathcal{Z}$ is associative but not invertible, becomes invertible, but is no more associative in general.

Consider now the functional h_r in $\mathcal{R}$ defined by

$$h_r(\mathbf{a}) = \rho(\alpha_1, \alpha_2) = H(\alpha_1|\alpha_2) + H(\alpha_2|\alpha_1)$$

It is almost immediate to proove that:

i) h_r coincides in $\mathcal{R}^\nu$ with the ordinary entropy H in $\mathcal{Z}$.

ii) $h_r(\mathbf{a}) \geq 0$, and $h_r(\mathbf{a}) = 0 \Rightarrow \mathbf{a} = \mathbf{u}$

iii) $h_r(\mathbf{a}^{-1}) = h_r(\mathbf{a})$

iv) if $\mathbf{a}$ and $\mathbf{b}$ are r.i. r-partitions, $h_r(\mathbf{ab}) \leq h_r(\mathbf{a}) + h_r(\mathbf{b})$.

It is therefore plausible to consider h_r an entropy in $\mathcal{R}$, even if it is not unique, in the sense that there exist other functionals with the same properties *i-iv*.

Provided that the reduction process (i.e. the projection from $\mathcal{Z} \times \mathcal{Z} \to \mathcal{R}$) let the essential difference between two partitions emerge, h_r gives a quantitative estimate of such a "antisimilarity".

The next step consists in applying the concepts introduced above to the quantitative estimate of complexity in Cellular Automata. We recall that a Cellular Automaton is a triple $\{K, L, f\}$ consisting of an alphabet K, a lattice L and a map $f : K^L \to K^L$, which, for a given configuration $a \in K^L$, gives a new configuration $a' = f(a) \in K^L$, where the value in each vertex in the lattice depends only on a finite neighbour. To every configuration a we associate a partition $\alpha = \phi(a)$ (of course, the correspondence has to be meaningful with respect to the exigences of a definite problem). For instance, let $\mathbf{M}$ be a finite subset of L, say a hypercube: then the collection of homogeneous connexe subsets of $\mathbf{M}$ (where vertices are homogeneous if labelled by the same symbol in K, and connexe if they may be joined with a homogeneous walk) is a partition of $\mathbf{M}$. Starting from a configuration a_0, f produces the sequence $a_{k+1} = f(a_k)$, and a transformation T_f in $\mathcal{Z}$ is therefore defined through the relation: $\alpha_{k+1} = T_f(\alpha_k) = \phi(f(a_k))$. Provided that $\mathbf{M}$ is sufficiently large, the problem of the complicacy of the automaton may be rephrased in the problem of giving a reasonable estimate of the creation of novelty in the sequence $\alpha_0, \alpha_1, \alpha_2, \ldots$ In other words, we ask how different is α_{k+1} from α_k. Let then be, more generally,

$$\mathbf{a}_{n,k} = \pi(\alpha_{k+n}, \alpha_k)$$

The reduction process π eliminates, as far as possible, the spurious similarity between the k-th and the $(k+n)$-th steps in the evolution, so that $\mathbf{a}_{n,k}$ reflects the vanishing of the memory after n steps. Clearly the case $n = 1$ is specially noteworthy, since refers to the antisimilarity created step by step. Seen as a function of "time" k, the quantity

$$I_M(\alpha, n, k) = h_r(\mathbf{a}_{n,k})$$

is the first and fundamental indicator we introduce for the analysis on the complexity of the automaton. It is quite obvious, for instance, that all the properties of periodicity are reflected in I_M, and may be examined with standard tools, such as the power spectrum of its Fourier transform. Another indicator is the time average of the previous one:

$$A_M(\alpha, n) = \lim_N \frac{1}{N} \sum_{k=1}^{N} I_M(\alpha, n, k)$$

Of course, for the characterization of the automaton the dependence on the initial configurations and **M** shall also be taken into account; but it is plausible that a "great" size of **M** is sufficient to get stable and meaningful informations on the limit behaviour. We stress moreover that the analysis of complexity in a dynamic process should reflect the kind of interest the observer has, and that a finite time of observation or a finite window **M** can be as much significant, in certain contexts, as the limit quantities. For one dimensional CA, both I_M and A_M are accessible to explicit calculations. Numerical experiments are in progress indeed, and first results seem to confirm the effectiveness of the procedure in pointing out also hidden qualitative features connected to the complexity of CA.

We only mention here two further points: 1) as a time average of the local production of antisimilarity, A_M may be seen as an analogous quantity to the Lyapunov exponents measuring the rate of divergenge of nearby orbits in the phase space of a dynamical system. In this spirit, it is possible to introduce other indicators making this analogy even sharper: for a given orbit of partitions $\{T^n\alpha\}$, it is sufficient indeed to compute the rate of divergence of a second near orbit β, which evolves freely for a short time and is then drawn near to the first orbit. In this way, the usual computing procedure for Lyapunov exponents is replicated in $\mathcal{Z}$ or $\mathcal{R}$ as faithfully as possible. 2) the same indicators introduced for CA may be easily adapted to a variety of discrete process (shifts, transformations of geometrical patterns, evolution of fractality in attractors etc.), the only delicate point consisting in a correct and meaningful definition of the map ϕ into a partition space. Numerical experiments are in progress on both these subjects.

References:

1) Wolfram, S., Physica **10D**, 1 (1984)

2) Sinai, Ya. G., "Introduction to Ergodic Theory", Princeton University Press, Princeton N.J. (1976)

ONE-DIMENSIONAL COLLAPSE

O.Bokhove and A.Compagner

Laboratory of Applied Physics
P.O.Box 5046, 2600 GA Delft, the Netherlands

ABSTRACT

Molecular-dynamics simulations of a one-dimensional system of purely attractive particles, interacting with a rectangular-well pair potential, are briefly reported. The relevance to certain fundamental problems in statistical mechanics and in cosmology is indicated.

INTRODUCTION

Self-gravitating systems such as spherical star clusters and galaxies are difficult to simulate and seem to defy a description by means of statistical mechanics. See Lynden-Bell[1] for a review. The gravitational potential $\phi(r) = -Gm^2/r$ between two masses m at a distance r poses three problems: it has a long range, it is singular in $r = 0$, and it is negative everywhere. Leaving the first two problems aside one may focus one's attention on the third difficulty, by studying the statistical-mechanical properties of a system of particles interacting through a non-singular but purely attractive pair potential of short range. The potential energy of such a system still violates the stability condition.

An approximate way to deal with these systems is given by a cell model due to Hertel and Thirring[2], recently revisited by Compagner et al.[3] in a modified version. In this model, the volume occupied by the system is divided into a large number of cells. Each pair of particles residing in the same cell contributes a negative amount $-\varepsilon$ to the potential energy, whereas particles in different cells do not interact at all. This cell model can be generalized in different ways, but it is useful to ask first whether the cell model does indeed give a reasonable description of the corresponding system governed by true pair potentials. Since particles with short-range attractions without hard core are rather hypothetical, only computer experiments can tell.

The cell model was found to agree remarkably well with molecular-dynamics results for two-dimensional systems with a smooth attractive pair potential of short range, but due to the long relaxation times that were encountered the systems that could be studied were rather small[3)]. Can this situation be improved by restricting the attention to one-dimensional systems?

MOLECULAR DYNAMICS

Consider a system of N identical particles in a one-dimensional 'volume' V, with total energy E. The particles interact through a rectangular-well pair potential, with depth ε and width σ, and without hard core. Instead of E, the characteristic parameter $\eta = 2E/\varepsilon N^2 - 1/N$ can be used ($\eta = -1$ corresponds with the completely collapsed state with zero kinetic energy, $\eta \gg 0$ with a homogeneous phase, and $\eta \approx 0$ with a transition region).

Simulation is not difficult. A particle moves freely until a collision takes place, that is, until its distance to any other particle is equal to the width σ of the potential well, when the respective velocities jump to other values. As a pilot experiment, a system of 60 particles in a volume $V = 75\ \sigma$ with periodic boundaries was followed until 5.10^5 collisions had taken place. The results are shown in Fig.1. Along a horizontal line in the figure the momentary positions of all particles are indicated by a dot, subsequent lines showing momentary configurations separated by 2000 elementary collisions. One of the particles is indicated by small open circle instead of a dot, but this is barely visible.

Fig.1(a) was obtained for $\eta = 0.026$, just at the cell-model threshold for partial collapse. Most of the time a rather small cluster of fluctuating size is present, but during the period indicated by the square bracket it dissolved almost completely. During that period the kinetic energy was observed to be low.

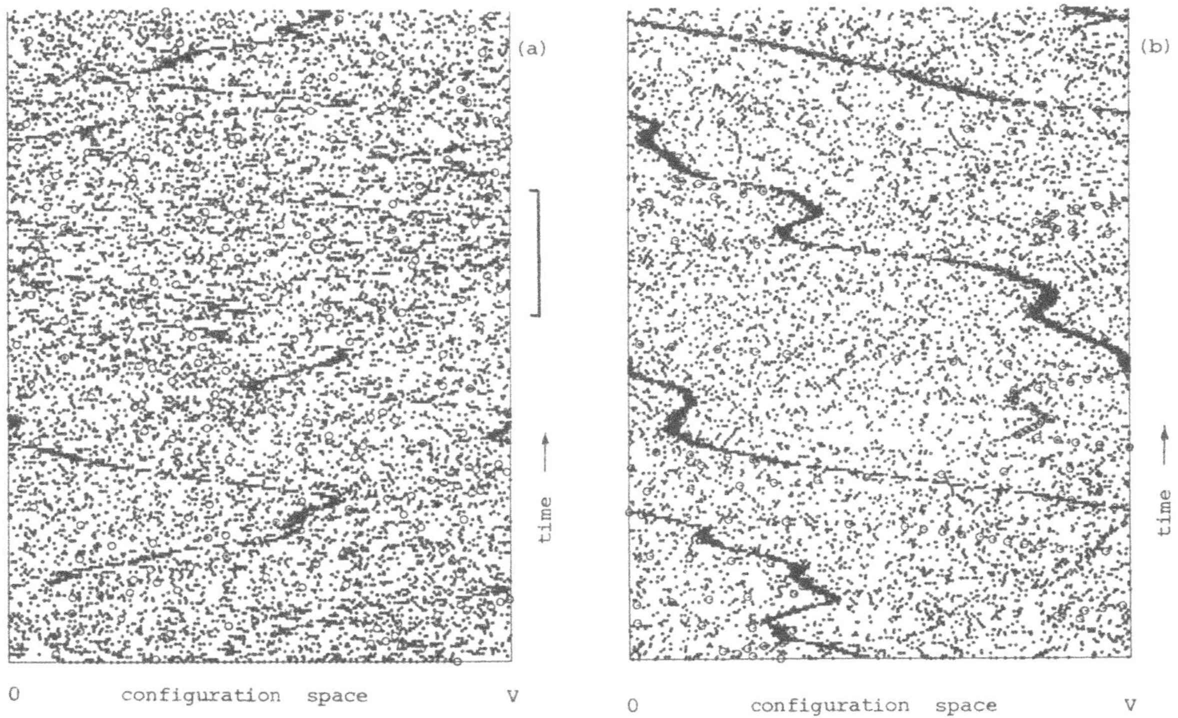

Fig.1. For a system of 60 particles, interacting through a rectangular-well potential, 250 momentary configurations separated by 2000 pair 'collisions' are shown. The 'volume' V corresponds with 75 well widths. (a) Just below the energy threshold for partial collapse, the system hesitates between a homogeneous phase and a phase with a small fluctuating cluster. (b) Well below the threshold a single stable cluster of, in this case, about 25 particles is present.

Fig.1(b) was obtained at $\eta = -0.1$. The slowly wandering cluster, containing about 25 particles, appeared to be rather stable.

Unfortunately, for both values of η the relaxation times turn out to be too long to obtain reliable values for the average cluster size, nor could temperature, pressure and specific heat (expected to be negative in a certain energy range) be calculated accurately. Longer simulations are in progress.

CONCLUSION

A first-order transition between a homogeneous phase at high energies and a collapsing phase with a single stable cluster in a homogenous background of reduced density was observed in the earlier simulations of two-dimensional systems of purely attractive particles[3]). The cell-model predicts the existence of this transition also for one-dimensional systems. The present computer experiments are not in conflict with this prediction.

REFERENCES

1) Lynden-Bell, D., in Gravitation in Astrophysics, Vol. 156 of NATO Advanced Study Institute, Series B: Physics, ed. by B.Carter and J.B.Hartle (Plenum, New York, 1987), p.155.
2) Hertel, P., and Thirring, W., Ann.Phys. 63, 520 (1971).
3) Compagner, A., Bruin, C., and Roelse, B., Phys.Rev. A39, 5989 (1989).

SIMULATION OF DISLOCATION GLIDE INCLUDING THERMAL ACTIVATION

D. RÖNNPAGEL and T.S.J. STREIT†
Institut für Metallphysik und Nukleare Festkörperphysik and
Institut für Theoretische Physik† TU Braunschweig.
Fed. Rep. of Germany

Abstract

We study the motion of dislocations due to external stresses and thermal activation through random arrays of spherical obstacles. Thermal equilibrium between the dislocations and the crystal is achieved by a stochastic term introduced into a partial differential equation which applies for the deterministic mechanical behaviour of the system. The results are compared with measurements on single crystal alloys.

Numerical simulation of dislocation glide through random arrays of obstacles has proved to be a fruitfull method in reproducing experimental results as well as in testing the theoretical models. In this work we present a new approach of taking into account thermal activation of plastic deformation. We will treat the influence of temperature during the overcoming of single obstacles and arrays of obstacles distributed randomly in three dimensions.

1 The Stochastic Equation of Motion of a Dislocation Line

We describe the dislocation line by $y(x,t)$. (x,y) ist the glide plane. The deterministic equation of motion of a dislocation was derived [1] from an energy balance being applied for each infinitesimal dislocation segment. This results in the condition that the sum of all stresses acting perpendicularly on a dislocation segment has to vanish.

$$\tau_{\text{inertia}} = \tau_{\text{back}} + \tau_{\text{drag}} + \tau_{\text{ext}} + \tau_{\text{obst}} \tag{1}$$

The shear stress τ_{ext} and the stresses due to the obstacles τ_{obst} are derived by means of the Peach Koehler formula

$$\begin{aligned} b \cdot \tau_{ext} &= \vec{b} \cdot \sigma_{ext} \cdot \vec{n}_0, \\ b \cdot \tau_{obst} &= -\vec{b} \cdot \sum_i \sigma_{obst}^{ys}(x, y, \vec{r}_i) \cdot \vec{n}_0 \end{aligned} \qquad (2)$$

($\vec{b} = b(0,1,0)$ denotes the Burgers vector of the dislocation, $\vec{n}_0 = (0,0,1)$ its glide plane normal): σ_{obst} represents the interaction between dislocation and glide obstacle. In this work we have used the stress tensor of elastic spheres embedded into spherical holes off slightly smaller radii being cut out off an elastic matrix:

$$\sigma_{obst}^{ys}(x, y, \vec{r}_i) = -\frac{A \cdot y_i \cdot z_i}{((x - x_i)^2 + (y - y_i)^2 + (z - z_i)^2 + w^2)^{5/2}} \qquad (3)$$

The parameters A (obstacle strength) and w (obstacle width) shall be adjusted by comparing the results of the simulation with experimental data.
τ_{back} is the stress which arises from dislocation curvature

$$\tau_{back} = 1.25\frac{\mu \cdot b}{4\pi}\frac{1}{R}\ln(c_2\frac{R}{b}) \simeq \frac{\mu \cdot b}{2}\frac{\partial^2}{\partial x^2}\, y(x,t), \; \frac{1}{R} = \frac{\partial^2 y}{\partial x^2}(1 + (\frac{\partial y}{\partial x})^2)^{-1.5}, \qquad (4)$$

where μ is the shear modulus, c_2 a constant near to 1.
τ_{drag} and $\tau_{inertia}$ are the friction stress (case of viscpus drag) and the inertial stress respectively. They are introduced acting perpendicularly on each segment. Denote $\dot{s}_n$, the normal velocity and $\ddot{s}_n$ the normal aceleration off the dislocation , B the drag coefficient and m the linear mass density, then:

$$\tau_{drag} = -\frac{B}{b}\dot{s}_n \simeq -\frac{B}{b}\frac{\partial}{\partial t}\, y(x,t), \qquad (5)$$

$$\tau_{inertia} = \frac{m}{b}\ddot{s}_n \simeq \frac{m}{b}\frac{\partial^2}{\partial t^2}\, y(x,t), \qquad (6)$$

After substituting Eq. (2)–(5) into Eq. (1) we introduce next the thermal stresses due to the interaction with the heat bath to get a stochastic differential equation of motion:

$$(\frac{m}{b}\frac{\partial^2}{\partial t^2} + \frac{B}{b}\frac{\partial}{\partial t} - \frac{\mu \cdot b}{2}\frac{\partial^2}{\partial x^2})\, y(x,t) = \tau_{ext} + \tau_{obst} + \frac{\Gamma(x,t)}{b}, \qquad (7)$$

with $\Gamma(x,t)$ a random variable in space and time. The thermal stresses have to be chosen in such a way that the states of the system are distributed with a probability density given by the Boltzmann distribution [2]:

$$\begin{aligned} < \Gamma(x,t) > &= 0, \\ < \Gamma(x_i,t)\Gamma(x_j,t') > &= q \cdot \delta(x_i - x_j) \cdot \delta(t - t');\; q = \frac{2 \cdot B}{m \cdot (kT)^{-1}} \end{aligned} \qquad (8)$$

One can show that the stochastic Langevin equation (7) corresponds to a Klein-Kramers equation of motion for the probability density of the states of the system, the stationary solution of which is the Boltzmann distribution.

2 Results

To compare with experiment the material constants are chosen to be the values for copper; $\mu = 4.21 \cdot 10^{10}\frac{N}{m^2}$, $b = 2.56 \cdot 10^{-10}m$, $B = 1.7 \cdot 10^{-5}\frac{Kg}{m \cdot sec}$, and $m = b \cdot 1.055 \cdot 10^{-6}\frac{Kg}{m^2}$. The obstacles are distributed randomly on the Cu fcc lattice with the glide plane centered between two (1,1,1) planes.

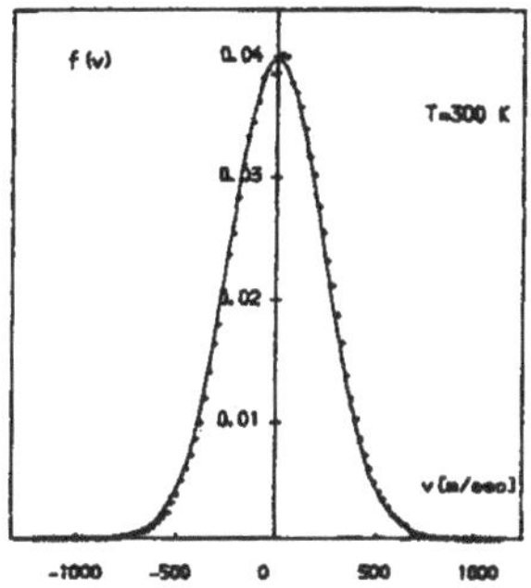

Fig. 1. Velocity distribution f(v) for segments of length 0.8b, for a dislocation trapped between obstacles, at 300K, with small external stress. The points are the simulation results after 10^6 numerical steps of integration. The full line shows the expected Maxwell-Boltzmann distribution.

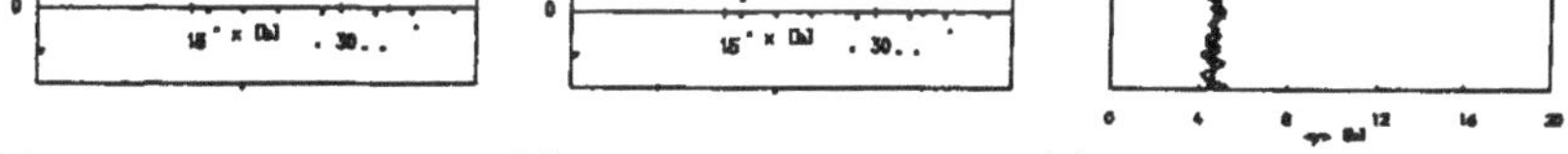

(a) (b) (c)

Fig. 2. Glide of the dislocation through an 3D array of obstacles (each represented with an x) under a slowly increasing linear external shear stress. For the initial configuration the external shear stress and the temperature are to low to overcome the obstacles. (a) shows the glide for T = 0K , (b) for T = 300K. and (c) shows τ_{ext} as function of the average position of the dislocation at various temperatures, thus giving the temperature dependence of the critical shear stress. The obstacles were chosen with a maximum τ_{obst} of 800 MPa, width 0.8b being placed randomly on the Cu fcc lattice with concentation c=2% at.

Comparison with the results of Fig. 2c) with experiment [3] show that the effect of the obstacles is to strong, so that we have to adjust the parameters of the obstacle stress.

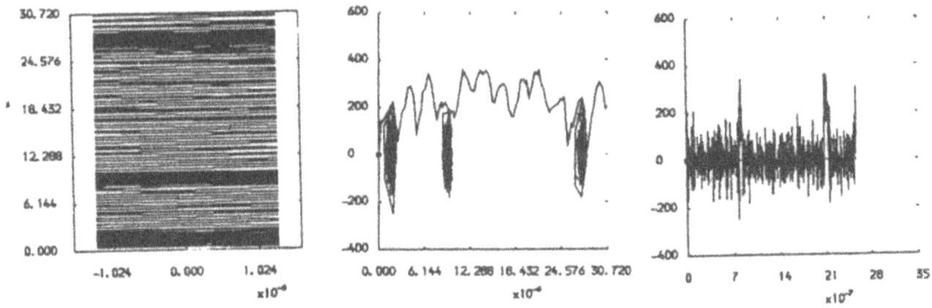

(a) (b) (c)

Fig. 3. Dislocation of length 10b passing through a periodic sequence of 12 obstacles. The obstacles are placed on the y-axis at 10b, 20b, ..., 120b. The external stress is $\tau_{ext} = 57$Mpa, the temperature 300K. (a) shows the motion in the (x,y) plane. The dislocation gets trapped three times. (b) shows the mean velocity as function of the average position of the dislocation. (c) gives the velocity as function of time. The time intervals where the dislocation is gliding (due to the external stress and its kinetic energy) are very small compared to the trapped ones where the dislocation has to overcome the obstacle with thermal activation exclusively. The resulting order of average velocity amounting to 1.710^{-3} m/sec is in agreement with experimental data found in load pulse experiments.

References

[1] D. Rönnpagel. Simulation Calculations of the Static and Dynamic Equilibrium of Dislocation Segments. In Constitutive Relations and their Physical Basis 8th Risø international Symposium on Metallurgy and Material Science , page 503, 1987.

[2] H. Risken. The Fokker-Planck Equation. Methods of solution and Applications. Springer-Verlag, Berlin Heidelberg New York Tokio, 1984.

[3] C. Schwink and T. Wille. Scripta Met. , 14:1093, 1980.

Conserved Order Parameter Dynamics and Chaos in One Dimensional Discrete Model +

S.Puri*, R. Kapral** and Rashmi C. Desai

Department of Physics, University of Toronto,

Toronto, Ontario, Canada M5S 1A7

ABSTRACT

We consider the discretised, one-dimensional, deterministic non-linear map that is the discrete analog of the Cahn-Hilliard model of phase separation. The role of conservation law and instability is explored. In contrast to the continuum problem, the discretisation leads to chaotic solutions. The bifurcation structure that emerges is discussed.

1. INTRODUCTION

The Cahn-Hilliard model[1] is the simplest example of the nonlinear dynamics for a phase separating system with a conserved order parameter[2]. In d-dimensional systems with $d > 1$, the interfaces are formed and evolve in time such that local interfacial curvature plays a crucial role in the nonequilibrium phenomena of phase separation. This complicating feature is absent in one dimen-

---

+ Research supported by the Natural Sciences and Engineering Research Council of Canada.

* Permanent address : School of Physical Sciences, Jawaharlal Nehru University, Vedanta Desika Marg, New Delhi – 110067, India.

** Department of Chemistry, University of Toronto, Toronto, Ontario, Canada M5S 1A1.

sional systems and it is therefore of interest to inquire the role that the nonlinearity, instability and conservation law play in the 1-dimensional phase separation dynamics.

In terms of dimensionless variables, the 1-d Cahn-Hilliard equation for a conserved order parameter field $\psi(x,t)$ is

$$\begin{aligned}\frac{\partial\psi(x,t)}{\partial t} &= -\frac{\partial^2}{\partial x^2}\left[\psi(x,t)-\psi(x,t)^3+\frac{\partial^2\psi(x,t)}{\partial x^2}\right],\\ &\equiv -\frac{\partial^2}{\partial x^2}\mu(x,t),\\ &\equiv -\frac{\partial}{\partial x}J(x,t)\end{aligned} \tag{1}$$

Here the second line serves to define (*negative of*) the local chemical potential $\mu(x,t)$ and the third line the local current $J(x,t)$. A simple Euler-like approximation to this equation yields a discrete space and time coupled map model :

$$\psi(i,t+1)=\psi(i,t)-\frac{\tau}{\alpha^2}\Big[\mu(i+1,t)+\mu(i-1,t)-2\mu(i,t)\Big], \tag{2}$$

where τ and α are time and space steps respectively. Using the explicit form of the chemical potential we may also write eq. (2) in the form :

$$\begin{aligned}\psi(i,t+1)=\psi(i,t)-\frac{\tau}{\alpha^2}\Big[&\psi(i+1,t)+\psi(i-1,t)-2\psi(i,t)\\ &-\Big(\psi(i+1,t)^3+\psi(i-1,t)^3-2\psi(i,t)^3\Big)\Big]\\ -\frac{\tau}{\alpha^4}\Big[\psi(i+2,t)-4\psi(i+1,t)&+6\psi(i,t)-4\psi(i-1,t)+\psi(i-2,t)\Big].\end{aligned} \tag{3}$$

For sufficiently small τ and α the solutions of eq. (3) can be made to approximate those of eq. (1); however, in many recent studies of the phase separation problem rather large space and time steps are used in order to improve the computational efficiency of the model. Provided care is taken in the selection of τ and α so that unphysical bifurcation boundaries are not crossed, the coupled map model reproduces the gross features, like scaling properties, of the phase separation process. The discrete nature of the coupled map model has important implications for the interfacial structure and dynamics. We examine the interfacial structure in detail below; the results related to the dynamics will be described elsewhere.

2. STATIC STRUCTURE

The stationary homogeneous and inhomogeneous solutions for the order parameter field $\psi(i)$ can be obtained from eq. (2) or eq. (3) by setting $\psi(i,t) = \psi(i)$. The resulting equation can be written as a four-dimensional conservative mapping if new variables are defined. The discrete analog of the derivative of the order parameter field is

$$\alpha y(i) = \psi(i) - \psi(i-1) , \tag{4}$$

while the discrete chemical potential can be defined as

$$\alpha\mu(i) = y(i+1) - y(i) + \alpha\psi(i) - \alpha\psi(i)^3 , \tag{5}$$

and its discrete flux is

$$\alpha J(i) = \mu(i+1) - \mu(i) , \tag{6}$$

In terms of these variables the conservative mapping in the space variable is:

$$\begin{aligned}
\psi(i+1) &= (1-\alpha^2)\psi(i) + \alpha^2\psi(i)^3 + \alpha y(i) + \alpha^2\mu(i) , \\
y(i+1) &= -\alpha\psi(i) + \alpha\psi(i)^3 + y(i) + \alpha\mu(i) , \\
\mu(i+1) &= \mu(i) + \alpha J(i) , \\
J(i+1) &= J(i) .
\end{aligned} \tag{7}$$

The last line of eq. (7) expresses the conservation of flux in the steady state. Letting $\mathbf{u}$ be the vector field $\mathbf{u}^T(i) = (\psi(i), y(i), \mu(i), J(i))$, where T stands for the transpose, we may write the mapping in compact form as

$$\mathbf{u}(i+1) = \mathbf{F}(\mathbf{u}(i)) , \tag{8}$$

where the vector-valued map $\mathbf{F}$ is defined by comparison of eq. (7) and eq. (8). The inverse mapping can also be easily written.

The fixed point solutions of the mapping eq. (7) correspond to the spatially homogeneous states of the system. Denoting the fixed point values by an asterisk we find $y^* = J^* = 0$ while the order parameter field is given by the solution of $\psi^{*3} - \psi^* + \mu^* = 0$. Note that the chemical potential $(-\mu^*)$ acts like an external field on the system and that as expected this fixed point equation has the same structure as that for the Ginzburg-Landau model for a nonconserved order parameter field[3).

The solution structure has the usual sigmoidal shape (Figure 1) in the ψ-μ-plane, and for $-\frac{2}{3\sqrt{3}} \leq \mu^* \leq \frac{2}{3\sqrt{3}}$ there are two temporally stable homogeneous states separated by an unstable homogeneous state. For $\mu^* = 0$ the two temporally stable states are equivalent and have the same stability.

3. MANIFOLD STRUCTURE

The structure of the stable and unstable manifolds associated with the fixed points of the map determine the nature of the stationary inhomogeneous solutions. The character of the fixed points can be determined in the usual manner from a linear stability analysis about the fixed points. Letting $\mathbf{u}(i) = \mathbf{u}^* + \mathbf{u}'(i)$ the linearized version of eq. (8) is:

$$\mathbf{u}'(i+1) = \mathbf{L}_F(\mathbf{u}^*)\mathbf{u}'(i) \equiv \left(\frac{\delta \mathbf{F}(\mathbf{u}(i))}{\delta \mathbf{u}(i)}\right)_{\mathbf{u}(i)=\mathbf{u}^*} \mathbf{u}'(i) \,. \tag{9}$$

The eigenvalues of $\mathbf{L}_F$ are $\lambda_1 = \lambda_2 = 1$, $\lambda_3 = \lambda_+$ and $\lambda_4 = \lambda_-$ where

$$\lambda_\pm = 1 - \frac{\alpha^2 - 3\alpha^2\psi^{*2}}{2} \pm \left[[1 - \frac{\alpha^2 - 3\alpha^2\psi^{*2}}{2}]^2 - 1\right]^{1/2} . \tag{10}$$

The temporally stable fixed points are hyperbolic since $\lambda_+ > 1$ and $0 < \lambda_- < 1$, while the temporally unstable fixed point is elliptic provided $\alpha^2(1 - 3\psi^{*2}) < 4$ and hyperbolic with reflection if $\alpha^2(1 - 3\psi^{*2}) > 4$. The marginally stable eigenvalues λ_1 and λ_2 arise as a result of the conservation law that is evident in eq. (1).

Figure 2 shows phase portraits of the map, eq. (7), obtained by iterating a set of initial conditions near to and in between the temporally stable fixed points. The initial flux was taken to be zero, $J(0) = 0$; hence, the chemical potential is constant and can be used to parameterize the dynamics which is confined to the ψy-plane. The hyperbolic and elliptic characters of the various fixed points are evident in these figures. Figure 2a in the upper panel shows the phase portrait for $\mu = 0$ and $\alpha = 1.0$. Compared to the expected results for the Cahn-Hilliard equation, the saddle connection is destroyed in the discrete model. The saddle area is shown under magnification in the lower panel. A complex manifold structure that gives rise to chaotic dynamics is obtained. Similar results obtain for $\mu \neq 0$ as can be seen in Fig. 2b and 2d. Of course, the external field μ destroys the saddle connection between the two different hyperbolic points. If α is decreased qualitatively similar results are also obtained in the saddle area upon appropriate magnification. Our results indicate that even a fine discrete space step destroys

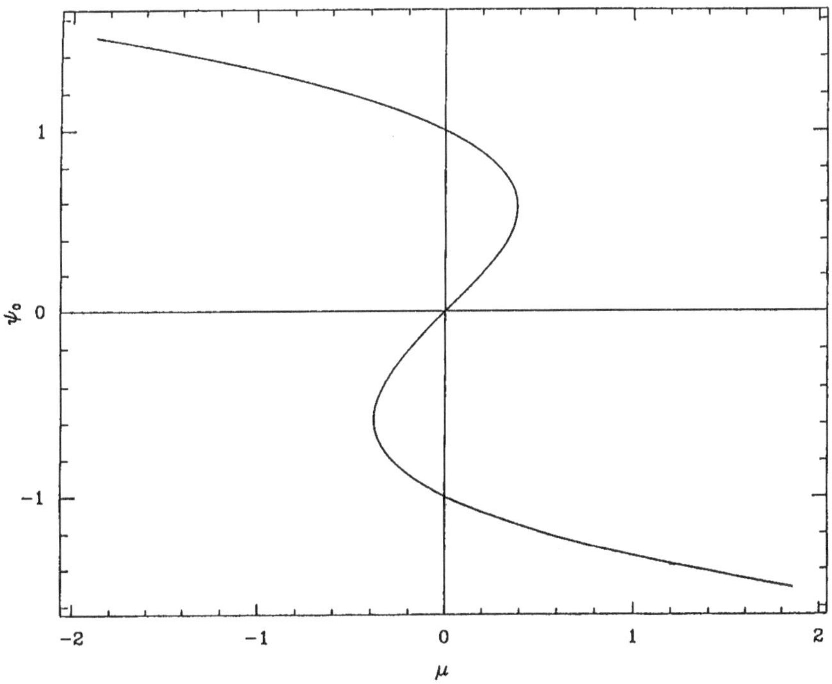

Figure 1 : Spatially homogeneous states $\psi_0(\mu)$ obtained from the fixed point solutions of the nonlinear map in eq.(7).

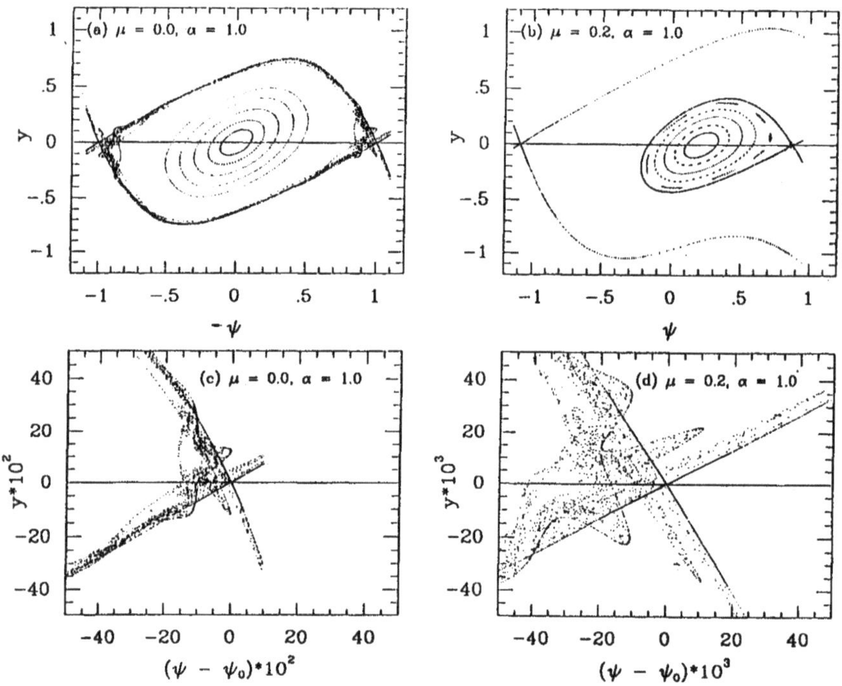

Figure 2 : Phase portraits of the map, eq.(7), displaying its manifold structure for the indicated values of μ and α.

the (nongeneric) saddle connection leading to a chaotic map in the spatial variable; however we find that well-defined inhomogeneous states are also obtained.

It is instructive to examine the manifold structure in more detail. There is only a single eigenvector corresponding to the doubly degenerate eigenvalue $\lambda = 1$ so that the matrix $\mathbf{L}_F$ is defective. The corresponding eigenvector is $\mathbf{e}_1^T = (1, 0, 1 - 3\psi^{*2}, 0)$, which represents a displacement along the line of fixed points in the $\psi\mu$-plane; such a displacement should not give rise to relaxation or growth. The eigenvectors corresponding to $\lambda_\pm$ are $\mathbf{e}_\pm = (1, (\lambda_\pm - 1)/(\lambda_\pm \alpha), 0, 0)$. Note that these vectors are confined to the ψy-plane; hence, if there is no initial component along μ and J the dynamics will be confined to the ψy-plane. The stable W_s and unstable W_u manifolds associated with the two hyperbolic fixed points (i=1,2) are fixed point functions under the map, eq. (8), and may be constructed in the usual manner by iterating a set of points along the directions of the eigenvectors $\mathbf{e}_\pm$ and slightly displaced from the fixed points. In this way we can construct stable $W_s^{(i)}$ and unstable $W_u^{(i)}$ manifolds associated with each hyperbolic point.

4. SPECIAL STATIC SOLUTIONS

The stationary inhomogeneous states of the system and interfacial profiles can be discussed in terms of the manifolds described above. Figure 3 shows the equilibrium manifold structure for $\mu = 0$ and $\alpha = 1.0$ (same as in Figure 2a), with additional information related to some special static solutions in terms of this manifold structure.

The simplest solution is the 'kink' that connects the two homogeneous states denoted by ψ_1^* and ψ_2^*. One can iterate map, eq.(7), from an initial state where half of the system was in ψ_1^* while the remainder was in ψ_2^*. From such a calculation we find that the kink solution must arise from the heteroclinic intersections of the stable and unstable manifolds of these two fixed points. The discrete profile corresponds to a trajectory in space that emanates from the fixed point ψ_1^* along the unstable manifold $W_u^{(1)}$ and enters ψ_2^* along the stable manifold $W_s^{(2)}$; thus, $W_u^{(1)} \cap W_s^{(2)}$ or $W_s^{(1)} \cap W_u^{(2)}$. Figure 3 shows the locations of the points (crosses) on the discrete interfacial profile superimposed on the manifold structure. As discussed in Ref. (3) these interfacial points lie on alternate heteroclinic interesctions of the stable and unstable manifolds. The origin of this alternating structure is connected with the temporal instability of the solution $\psi^* = 0$.

The discrete model also supports a variety of other stationary inhomogeneous states. We have explored a number of single bump solutions with different widths;

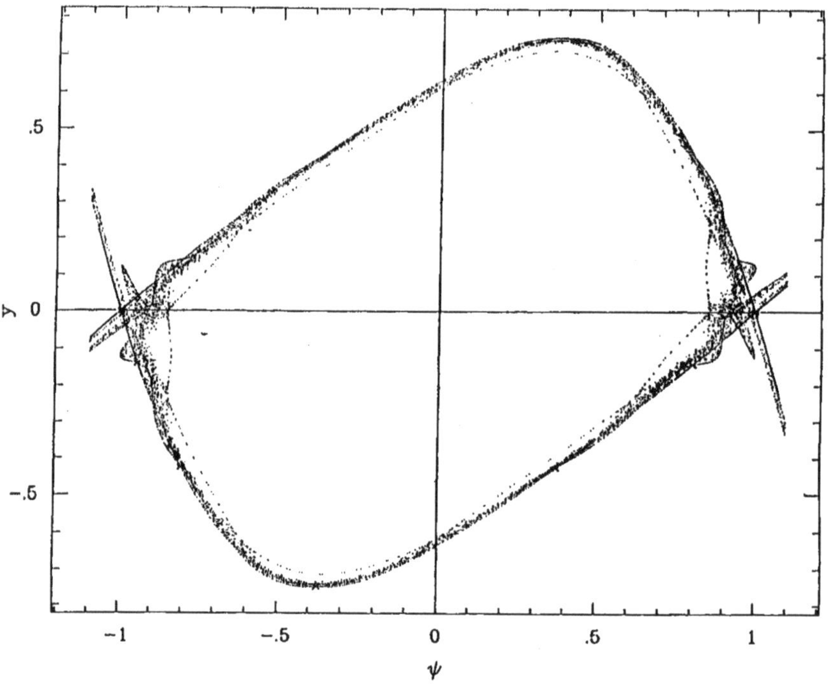

Figure 3 : Discrete interfacial profile (crosses) superposed over the manifold structure for $\mu = 0$ and $\alpha = 1$.

these are obtained by iterating eq. (7) from a square wave initial state of variable width where the background was ψ_1^* and the height of the square wave was ψ_2^*. The smallest width of a stable bump is a function of α. Since these solutions correspond to trajectories in space that leave the ψ_1^* fixed point and later reenter it, the discrete points in the bump profile correspond to homoclinic intersections of the stable and unstable manifolds associated with a single fixed point; e.g., $W_u^{(1)} \cap W_s^{(1)}$. We have also investigated the temporal stability of various stationary states discussed above. The details of these studies will be given elsewhere[4].

REFERENCES

1) Cahn, J.W. and Hilliard, J.E., J. Chem. Phys. 28, 258 (1958); Novick-Cohen, A. and Segel, L.A., Physica 10D, 277 (1984); Novick-Cohen, A., J. Stat. Phys. 38, 707 (1985).

2) Rogers, T.M., Elder, K.R. and Desai, R.C., Phys. Rev. B37, 9638 (1988); Gunton, J.D., San Miguel, M. and Sahni, P.S. in *Phase Transition and Critical Phenomena*, Ed. C. Domb and J.L. Lebowitz (Academic, New York, 1983), vol. 8, p. 267.

3) Oppo, G.-L. and Kapral, R., Phys. Rev. A36, 5820 (1987) and references therein.

4) Puri, S., Kapral, R. and Desai, R.C., Unpublished.

www.ingramcontent.com/pod-product-compliance
Lightning Source LLC
LaVergne TN
LVHW071223011026
849280LV00007B/42